FROM WATER SCARCITY TO SUSTAINABLE WATER USE
IN THE WEST BANK, PALESTINE

From Water Scarcity to Sustainable Water Use in the West Bank, Palestine

DISSERTATION
Submitted in fulfillment of the requirements of
the Board for Doctorates of Delft University of Technology
and the Academic Board of the UNESCO-IHE Institute for Water Education
for the Degree of DOCTOR
to be defended in public
on Thursday, November 26, 2009 at 15:00 hours
in Delft, The Netherlands

by

DIMA WADI' H. AL-NAZER

born in Hebron, Palestine

Master of Science in Water Engineering

Birzeit University, Palestine

This dissertation has been approved by the supervisors
Prof. dr. ir. P. van der Zaag
Prof. dr. H.J. Gijzen
Dr. M.A. Siebel, PE.

Members of the Examination Committee:
Rector Magnificus

Prof. dr. H.J. Gijzen	Vice-Chairman, UNESCO-IHE
Prof. dr. ir. P. van der Zaag	UNESCO-IHE and TU Delft, promotor
Prof. dr. D. Huisingh	University of Tennessee
Prof. dr.-ing. M. Jekel	TU Berlin
Prof. dr. ir. H.H.G. Savenije	TU Delft
Dr. Z. Mimi	Associate Professor Birzeit University
Dr. M.A. Siebel, PE.	Associate Professor UNESCO-IHE
Prof. dr. ir. N.C. van de Giesen	TU Delft

CRC Press/Balkema is an imprint of the Taylor & Francis Group, an informa business

Published by:
CRC Press/Balkema
PO Box 447, 2300 AK Leiden, The Netherlands
e-mail: Pub.NL@taylorandfrancis.com
www.crcpress.com – www.taylorandfrancis.co.uk – www.balkema.nl
ISBN 978-0-415-57381-8 (Taylor & Francis Group)

Dedication

To the soul of my father Wadi' Nazer, mother Adla Hidmi and brother Hussein Nazer.

Father, I still remember your words, for a friend of my sister who was invited for a meal in our house. The young gentleman had left some of the food in his plate after the meal. You said then "look son, if you knew how much money and effort had been spent to produce these leftovers, you will never leave anything in your plate anymore. For example, producing rice costs a lot of effort and resources on the country of production, it has been transported to our country and finally cooked to be ready in your plate. By disposing this amount you are neglecting all these efforts and resources. Moreover, you are spending part of your money for nothing". That was a lesson on life cycle and global thinking.

Mother, although more than 25 years passed the message is still valid. Once you bought a piece of cloth, it was small. You decided to make me a suit (jacket and skirt) from this cloth. I told you it is not enough to make two pieces out of it, let us do the jacket only. You insisted and put the cloth on the table and said: "let's find out the best way of organizing the patrons of the jacket and skirt on the cloth?" The job was rather difficult. However, after a while of thinking, you did it and got the two pieces out of the small cloth. That was my first lesson on managing scarce resources.

Hussein, you spent four years of you short life in the Israeli prisons because you believed in resisting the occupation for the sake of liberating man and land. You passed away without seeing that day coming. I will say today that man, land and water need to be liberated.

Dima Nazer
March 2009

Acknowledgments

Special appreciation is extended to all those who contributed their time and energy in any way in order to help in getting this mission finished.

It would have not been possible for me to undertake this study without the financial support from the Saudi Arabian Government, Ministry of Higher Education in the West Bank, Palestine and the NUFFIC fellowship program in the Netherlands. Their generous support is highly appreciated.

Special thanks are extended to my promoters Prof. Pieter van der Zaag from UNESCO-IHE for his assistance, helpful comments, fruitful discussions and encouragement. I appreciate his way of working with me; it made me feel his colleague rather than his student and Prof. Huub Gijzen from UNESCO, Jakarta Office for his advice and useful comments.

I would like to express my deep gratitude to Prof. Maarten Siebel for his unique cooperation, support, valuable comments, and advice. He was very patient, inspiring and encouraging through the whole period of my study.

Thanks are due to Prof. Ziad Mimi director of the Water Studies Institute, Birzeit University for his support, fruitful comments and discussions. Thanks are also extended to Dr. Amaury Tilmant for his help in the agriculture part of the research and Eng. Hisham Kuhail for his help in preparing the final version of the document.

Thanks are extended to the administrative staff of UNESCO-IHE for their help during my stay in the Netherlands. Thanks also to the administrative staff of Ministry of Higher Education in Palestine and of Palestine Technical Colleges-Arroub for their support, and in particular to Eng. Bassem Qumseiha, dean of the college.

I am deeply indebted to my husband Bassem Mokhtar for his forbearance, and patience during the entire period of the study. Bassem, without your support and encouragement this job would have not been finished or even started. You are always cooperating and understanding. Throughout this long study I came upon several difficulties, you were always there to support.

Thanks to my sisters Sahar, Samar, Layana, brother Safwan, and their families for their continuous encouragement and their support to my family during my absence.

At the early stages of my study, I was wondering about the wisdom of running for a PhD and leaving my family behind for long periods and bearing the vast amount of criticism for that. Samar, it was your encouragement that drove me forward when you said "go ahead, you may be criticized now, but your achievement will be admired later".

Last, but not least my deep thanks to my children Marwan, Muhannad, Maha and Saleem for their dedicated patience during my study. They are the ones who have sacrificed the most as a result of this work; they spent long times growing up without the mother when needed, they were children and they are now adults.

Table of contents

Abstract

The West Bank in Palestine is situated in the central highlands of Palestine. The area is bordered by the Jordan River and the Dead Sea in the east and the 1948 cease-fire line in the north, west and south and covers a total area of 5,800 km^2. It is a semi-arid area with limited water resources; the main water resource for Palestinians in the West Bank is groundwater. The per capita water availability, equal to the water use for all purposes, is 50 m^3/cap/year according to which the West Bank is considered an extremely water scarce area. Although the Palestinians can hardly meet their needs now, the situation will be even worse in the near future because of the expected increase in the population and developments in the social, commercial, industrial and environmental sectors.

The political situation in the area is making the water issue in the West Bank even more complicated. Since 1967, when Israel occupied the West Bank, the water resources are controlled by Israeli military orders which have severely restricted the Palestinian use thereof. The Palestinians are denied access to River Jordan Water. Moreover, there is inequitable distribution of the water resources in the area between Palestinians and Israelis; the per capita water use, for all purposes, of the Israelis is six times that of the Palestinians. In addition, the future water allocation between Palestinians and Israelis is unclear.

In addition to the politics-driven water scarcity situation, the common way of dealing with water in the West Bank does not help either. Water is used and disposed off without considering further uses. In most cases the used-water, or what is called wastewater, is discharged into the *wadis* (dry riverbeds) without any type of treatment, reducing water quality and, therefore, reducing availability of good quality water. Distribution systems are leaking and where water is available consumption figures are high at the expense of areas lacking water services.

The present study is based on the premise that the existing condition of Israeli control of the Palestinian water resources will continue during the projected study period through 2025.This is the worst case scenario. However, if more water would become available, the increased availability of water is expected to ease the water scarcity situation.

In short, the West Bank suffers from extreme water scarcity, has (for political reasons) less water than is naturally available, follows the "use and dispose" approach and anticipates an increase in the demand for water for reasons of population and economic growth. Therefore, the West Bank urgently needs a radical shift away from the present approach to water to one in which water is looked at as a scarce resource, in need of careful management so as to arrive at a situation where domestic, agricultural and industrial needs are satisfied within the limited water resources available and the environmental impact of used-water is significantly reduced.

The goal of this research is to develop a framework for the sustainable management of water resources in the West Bank. The approach in this thesis is based on the application of Cleaner Production thinking to water management.

To achieve this goal, three objectives were defined. The first objective is to prepare an inventory of the existing water use in the West Bank by determining its water footprint. The second objective is to evaluate options for water management, suitable for the Palestinian social, cultural, religious and economic conditions, now and in the near future, so as to reach water sustainability by 2025. In this context, three case studies demonstrating the feasibility of appropriate water management at the domestic, industrial and agricultural levels were developed. The third objective is to develop a water management strategy for the West Bank that aims at achieving water sustainability by 2025.

A prognosis of the water use in the West Bank was established by preparing a water balance and by calculating the water footprint of the Palestinians in the West Bank. It was found that the consumption component of the water footprint was 1,116 m^3/cap/year, compared to the global average of 1,243 m^3/cap/year. Local water resources provided only 50 m^3/cap/year out of which 16 m^3/cap/year was used for domestic purposes. This later number is only 28% of the global average and 21% of the Israeli domestic water use. Knowing these facts emphasizes the need for the above mentioned shift in thinking to an approach of *use, treat and reuse* instead of the common approach of *use and dispose*.

The domestic case study investigated options (rainwater harvesting, dual flush toilets, dry toilets...etc.) for improved domestic water management. These options were financially, environmentally and socially evaluated using Life Cycle Impact Assessment. The main conclusion was that by introducing a combination of water management options in the domestic area, a decrease in the water consumption of up to 50% can be achieved, thereby reducing the pressure on the scarce water resources. In addition to this environmental gain the financial impacts are being reduced. In the social context, it was found that introducing such options can improve the quality of life of those who presently do not have access to sufficient water. In fact, the *house of tomorrow* can be largely independent in terms of water and sanitation.

The second case study, related to irrigation in the West Bank, aimed at optimizing irrigation water use by using a linear mathematical model. Three scenarios were analyzed: The first scenario presents the existing cropping patterns, the second scenario maximizes profit under water and land availability constraints and the third scenario maximizes profit under constraints of water and land availability and local crops consumption. Results of the study showed that by determining the optimal patterns of the five crops included in the study, under land and water availability constraints, reduced the total agricultural water use by 4% while it can increase the profit in the entire agricultural sector by some 4%. It was concluded that water scarcity can be approached by changing the cropping patterns according to their water use. Moreover, expansion of rain-fed agriculture is key to planning the cropping patterns in water scarce countries.

The industrial case investigated the option of saving water and reducing pollution in the unhairing-liming process of the leather tanning industry. The conclusion was that the industrial process effluent could be reused after receiving the appropriate treatment. By doing so a substantial, up to 58%, reduction in water use can be achieved combined with a reduction in financial and environmental impacts.

Finally a strategy for sustainable water management in the West Bank was developed. The existing situation of the water sector was analyzed in terms of available resources, water use, Palestinian water rights, the National Water Plan, the institutional and organizational structure of the water sector as well as the expected availability and demand projections through the year 2025. In this context three scenarios were discussed;
1. The "do-nothing" scenario which assumes that the existing water availability will encounter no change due to the existing political situation (Israelis control over Palestinian water resources) while the population is increasing, thereby increasing the water demand.
2. The "water stress" scenario assumes that the overall water availability will increase following successful negotiations between Palestinians and Israelis. However, population growth and the development and improvements in the social, commercial, industrial and environmental sectors will increase the demand for water.
3. The "sustainable water use" scenario proposes a strategy for the sustainable water management. This last scenario was developed by using the results of a SWOT analysis. It includes technical and institutional improvements, the required legislation and regulations to support these improvements, the needed education for those who are going to implement or use them as well as the necessary economic incentives.

Under both the "do-nothing" and "water stress" scenarios there is an increasing gap between water availability and water demand. However, the proposed strategy in the sustainable water use scenario showed that this gap can be closed by gradually introducing water management alternatives that increase the availability (through rain-water harvesting) and reduce the demand through water conservation as well as re-use options.

The proposed alternatives in the industrial sectors proved to be financially feasible on the basis of the existing water price. In the domestic sector the proposed methods were found financially infeasible because of the high investment required for the new interventions. However, these investments become financially attractive when considering the social and economic benefits from improved health and social life which were not included in the calculations. WHO estimated these benefits at a global average of 8.1 US$ per dollar invested. Besides these investments are supposed to be paid by people who are going to benefit from these improvements and whose willingness to pay will be driven by the increasing water scarcity. Setting a reasonable pricing system that reflects the water scarcity will motivate people to invest in these improvements. Moreover, the international community may contribute to realizing the needed infrastructure.

With the present low water prices in the agricultural sector, the proposed approach to agricultural water saving is financially unattractive, also because of the high investments needed. However, constructing a reasonable water pricing system that ensures cost recovery in the agricultural sector will motivate farmers to use treated used-water for irrigation.

Implementing the combination of water management measures as proposed in this thesis will put the water management in the West Bank in Palestine on the sustainability track.

Chapter 1

Introduction

Introduction

1.1 Historical Background

Historically, Palestine is bordered by Lebanon in the north, Syria and Jordan in the east, the Mediterranean Sea in the west and Egypt and the Gulf of Aqaba in the south (Figure 1.1).

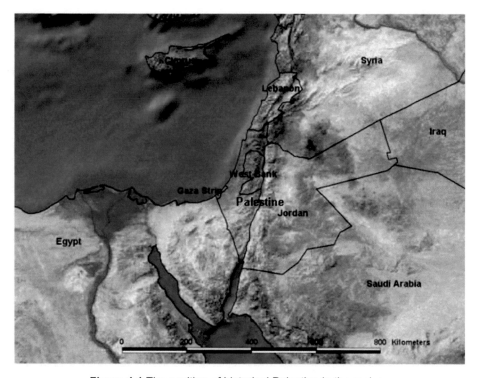

Figure 1.1 The position of historical Palestine in the region

Palestine was ruled by Arab Moslems between the seventh Century and the thirteenth Century. In 1099 the Crusaders attacked Jerusalem and later spread their control over the whole of Palestine until 1187 when Salah el Din al Ayuby defeated the Crusader army in Hiteen near Jerusalem. In 1291 the Turkish Mamluks entered Palestine and expelled the Crusaders from Palestine (ARIJ, 2000).

The Mamluks ruled Palestine until 1516 when the Othoman Turkish army invaded Palestine and ruled the area until World War I in 1917. British forces occupied Palestine in October 1917. On November 2nd, 1917, the British responded to Zionist demands through Arthur Balfour, the British foreign secretary, by declaring their support for creating a Jewish homeland in Palestine through the "Balfour Declaration" (ARIJ, 2000).

The British ruled Palestine under mandate until 1948 when the mandate was terminated and the Jewish State of Israel was declared within the land of Palestine by UN resolution 181. The Arabs rejected this resolution and did not announce a Palestinian state in the remaining land of Palestine. Therefore, the West Bank was ruled by Jordan and the Gaza Strip by Egypt (Figure 1.2). On June 5th, 1967 Israel occupied the West Bank and the Gaza Strip. The Palestinian non-violent resistance to occupation continued until the first "*Intifada*" - meaning "shaking up" in Arabic - in 1987 when the Palestinians used stones to attack the Israeli military forces.

On September 13, 1993, the Declaration of Principles (DOP) was signed by the Palestinian Liberation Organization (PLO), the representative body of the Palestinian people, and Israel. These principles initiated the implementation of the various peace agreements. The first interim agreement, known as Oslo I or Gaza-Jericho First, signed on May 4, 1994, was implemented and was followed by Oslo II in September 28, 1995 which was partially implemented because the Israeli-Palestinian negotiations encountered serious obstacles.

A protocol on redeployment from Hebron was renegotiated in January 17, 1997 and again partially implemented when a new Israeli government was elected. In October 22, 1998 a new memorandum was signed at Wye River and its first phase was implemented. Following the changes in Israeli government a new memorandum was signed at Sharm Al Sheikh in September 13, 1999 that aimed at implementing the modified Wye River memorandum (ARIJ, 2000). Due to the continuous delay in implementing the agreements, the second *Intifada* began on the 29th of September 2000. Since then, the area experienced high levels of violence. The situation became even more complicated after the election of "Hamas" in 2006 when Israel rejected the results of the democratic elections and increased the pressure on Palestinians by intensifying the closure to the extent of restricting food and fuel provision.

1.2 study area

This study focuses on the West Bank in Palestine, The West Bank is situated on the central highlands of Palestine; the area is bordered by the Jordan River and the Dead Sea in the east and the 1948 cease-fire line in the north, west and south. The total area of the West Bank is 5,800 km^2 including the area of the Dead Sea that falls within the West Bank boundaries (WRAP, 1994) (Figure1.2). The results of the 2007 census indicated that in December 2007 the total Palestinian population living in the West Bank was 2.4 million (PCBS, 2008). The population growth predictions on the basis of the 2007 survey are not available yet. However, the population predictions on the basis of 1997 census indicated that the projected population of the West Bank in 2025 will be 4.4 million. Three series of population projections were made; low, medium and high. The medium population forecast was used in this thesis. This projection assumes that the growth rate will decline from 3.8%/a to 2%/a during the period to 2025 (Table 1.1) (PCBS, 1999).

Table 1.1 Projected population in the West Bank during the period 1997 to 2025 (PCBS, 1999)

Year	2010	2015	2020	2025
Population*	3.0	3.5	3.9	4.4
*Population in millions				

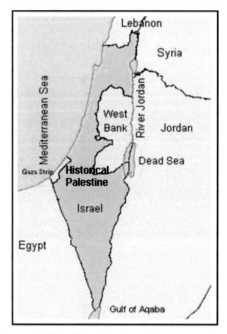

Figure 1.2 The West Bank and its surroundings

1.3 Water resources

Groundwater is the main water resource for Palestinians in the West Bank. The Jordan River is the only source of surface water in the area, to which the Palestinians have no direct access because Israel controls the flow of water from the river. Rainwater harvesting in cisterns forms an additional water resource for Palestinians in the West Bank (MOPIC, 1998 a).

The water quality in the West Bank is considered acceptable in general; there are no serious indications of pollution in the deep aquifers. However, there is contamination of water in the shallow aquifer wells and springs, which show indications of pollution with high levels of NO_3^- and Cl^- (MOPIC, 1998 c; ARIJ, 2001).

The water resources in the West Bank in Palestine are limited. The expected increase in the Palestinian population and development in social, commercial, industrial and environmental sectors will increase the pressure on the already scarce water resources.

Over-use of this resource is a serious threat and is the result of the complex Israeli-Palestinian political water issue and inadequate use of the scarce water resources (PNA, 1999).

The increasing consumption of water will result in an increase of wastewater production. The five public wastewater treatment plants in the West Bank are largely malfunctioning. In most cases the wastewater is discharged into wadis without any type of treatment, increasing the environmental problems (MOPIC, 1998 b; ADA and ADC, 2007). Moreover, increasing the percentage of the population connected to the sewer system - a standard policy approach, also in the West Bank - only increases the environmental deterioration when treatment capacity is not increased proportionally. Therefore, a new approach is needed to solve both the problem of water scarcity and of environmental deterioration. An approach that focuses on water conservation at the domestic, industrial and agricultural levels, ensures public health and improves environmental performance, an approach that minimizes the generation of waste, that treats the waste or even benefits from it through reuse or recovery of certain components.

The key for Palestinians is to manage their limited water resources adequately to ensure water sufficiency for future generations while improving the environmental situation.

The goal of the study is to develop a framework for the sustainable management of water resources in the West Bank in Palestine. The approach is based on the application of Cleaner Production thinking to water management

1.4 Objectives
The objectives of the study are
1. To prepare a water balance and an inventory analysis of the water use in the West Bank in Palestine by determining its water footprint.
2. To evaluate options for water use, suitable for Palestinian social, cultural, religious and economic conditions, now and in the near future, so as to reach water sustainability by 2025. This will be accomplished by developing case studies that will demonstrate the feasibility of appropriate water management at the domestic, industrial and agricultural levels.
3. To develop a water management strategy for the West Bank in Palestine that aims at achieving water sustainability by 2025.

1.5 Structure of the thesis
Figure 1.3 presents the structure of the thesis. Chapter 2 provides an inventory of the existing water resources in the area and a calculation of the water footprint for Palestinians in the West Bank. In chapter 3, an overview of the potential water management options in the domestic sector, suitable for the Palestinian society, is presented. A case study in the agricultural sector was developed in chapter 4. The objective of this study was to find the optimal cropping patterns in the West Bank in order to reduce water use for irrigation while at the same time maximizing profits. Chapter 5 presents an industrial case study aiming at reducing water consumption, environmental

impact and production cost of the unhairing-liming process in the leather tanning industry. In chapter 6 a water management strategy is presented that shows the direction in which the water sector in the West Bank is to be developed in order to achieve water sustainability by 2025. Finally, conclusions of the thesis are presented in chapter 7.

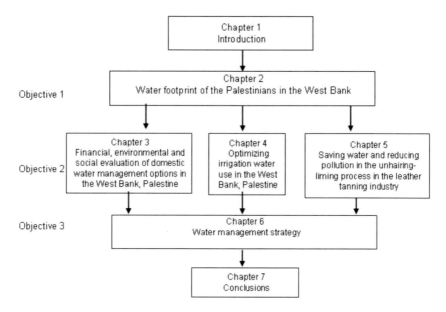

Figure 1.3 Structure of the thesis

References

Austrian Development Agency (ADA) and Austrian Development Cooperation (ADC) (2007) Water Sector Review, West Bank and Gaza, Volume I- summary Report (Final Report), prepared by Jansen and Consulting Team water consultant to the Austrian Development Agency for Palestine/Israel/Jordan, Jerusalem.

ARIJ (2000) An atlas of Palestine, West Bank and Gaza Strip, Applied research Institute of Jerusalem, Palestine

ARIJ (2001) Localizing Agenda 21 in Palestine, Applied research Institute of Jerusalem, Palestine

MOPIC a (1998) Emergency Natural Resources Protection Plan for Palestine "West Bank Governorates", Ministry of Planning and International Cooperation, Palestine.

MOPIC b (1998) Regional Plan for the West Bank Governorates, Water and Wastewater Existing Situation 1st ed., Ministry of Planning and International Cooperation, Palestine.

MOPIC c (1998) National Policies for Physical Development, Ministry of Planning and International Cooperation, Palestine.

Oslo II Agreement (1995), Israeli-Palestinian Interim Agreement on the West Bank and the Gaza Strip, Annex III, Article 40, Washington D.C., September 28 1995.

PCBS (Palestinian Central Bureau of Statistics) (1999) Population in the Palestinian Territory 1997-2025 , Palestinian Central Bureau of Statistics, Palestine.

PCBS (Palestinian Central Bureau of Statistics) (2008), Population, Housing and Establishment Census 2007, Census Final Results in the West Bank, Summary (Population and Housing), Ramallah- Palestine. available on line, http://www.pcbs.org

PNA (1999) Palestinian Environmental strategy, Main report, Ministry of Environmental Affairs, Palestinian National Authority (PNA), Palestine.

WRAP (1994) Palestinian Water Resources, A Rapid Interdisciplinary Sector Review and Issues Paper, The Task Force of the Water Resources Action Program, Palestine.

Chapter 2

Water Footprint of the Palestinians in the West Bank

Previously Published as

Nazer, D.W., Siebel, M.A., Mimi,Z., Van der Zaag, P. and Gijzen, H.J. (2008), Water footprint of the Palestinians in the west Bank, Palestine, Journal of American Water Resources Association (JAWRA), volume 44, issue 2, pp 449-458.

Water Footprint of the Palestinians in the West Bank

Abstract

Water in the West Bank of Palestine is a key issue due to its limited availability. Water is used from own sources for domestic, industrial and agricultural purposes. Moreover, water is consumed in its virtual form through consumption of imported goods, such as crops and livestock, the production of which used water in the country of production. In addition, wastewater in many parts of the West Bank is disposed off without treatment into the *wadis*, deteriorating the quality of the water resources in the area and, therefore, further reducing the quantity of good quality water available.

This paper calculates the water footprint for the West Bank. The consumption component of the water footprint of the West Bank was found to be 2791 million m^3/year. Approximately 52% of this is virtual water consumed through imported goods. The West Bank per capita consumption component of the water footprint was found to be 1,116 m^3/cap/year while the global average is 1,243 m^3/cap/year. Out of this number 50 m^3/cap/year was withdrawn from water resources available in the area. Only 16 m^3/cap/year (1.4%) was used for domestic purposes. This number is extremely low and only 28% of the global average and 21% of the Israeli domestic water use.

The contamination component of the water footprint was not quantified but was believed to be many times larger than the consumption component.

According to the definition of water scarcity, the West Bank is suffering from a severe water scarcity. Therefore, there is a need for a completely new approach towards water management in the West Bank, whereby return flows are viewed as a resource and that is geared towards a conservation oriented approach of 'use, treat and reuse'.

Key words: West Bank, water use, virtual water, water footprint, water scarcity.

2.1 Introduction

The water resources in the West Bank in Palestine are limited. There is water shortage in the area and this is expected to be more serious in the near future as both the population and the per capita consumption are increasing (MOPIC, 1998 a). Moreover, the water resources are threatened by water pollution due to the inadequate wastewater disposal which further decreases water quality and, therefore, availability.

Adequate management of water resources is important, specifically when resources are limited. A starting point for the adequate management of water is knowledge about the availability of water for the population and its economic activity. One way of expressing water use is through the concept of water footprint. The objective of this paper is to determine the water footprint for the West Bank.

2.2 Background
2.2.1 Water resources

Groundwater is the main source of fresh water in Palestine. Groundwater in the aquifer system flows in three main directions, according to which three main groundwater drainage basins can be, identified: the Western, the Northeastern and the Eastern basins.

The first two basins are shared between the West Bank and Israel, the eastern basin falls entirely within the West Bank (WRAP, 1994; MOPIC, 1998 a; SUSMAQ and PWA, 2001).

Surface water is considered to be of minor importance in the West Bank. The only source of surface water in the area is the Jordan River; Palestinian access to fresh surface water from the Jordan River is zero because the Israelis control the flow of the river (WRAP, 1994; MOPIC, 1998 a; ARIJ, 1998).Rainwater harvesting forms an additional source of water for domestic consumption in the West Bank. People collect rainwater falling on roofs or rock catchments and store it in cisterns, to meet part of their household needs (WRAP 1994; MOPIC, 1998 a). MOPIC (1998 b) estimated the quantity of harvested water in the West Bank at 6.6 million m^3/year.

2.2.2 Virtual water and water footprint

A good can be produced locally or can be imported. In the first case the production of the good requires the use of local water, in the second case the water is used in the country from where the good is imported. By consuming imported goods water is consumed in its virtual form. Virtual water is the water embodied in a good, not only in the real, physical sense, but mostly in the virtual sense. It refers to the water required for the production of a certain good (Allan, 1997).

To assess the water use in a country, we usually add up the water withdrawal for the different sectors of the economy. This does not give the real picture about the water actually needed by the people of that country, as many goods consumed by the people of the country are produced in other countries using water from that country (Hoekstra and Chapagain, 2007).

In order to have a consumption-based indicator of water use, the water footprint concept was developed by Hoekstra and Hung (2002) in analogy to the ecological footprint concept. The 'ecological footprint' of a population represents the area of productive land and aquatic ecosystems required to produce the resources used, and to assimilate the wastes produced by a certain population at a specified material standard of living, wherever on earth that land may be located (Wackernagel and Rees, 1996; Wackernagel *et al*, 1997; Wackernagel and Jonathan, 2001 cited in Chapagain and Hoekstra, 2004 a). The water footprint of an individual, business or nation then was the total annual volume of freshwater that is used to produce the goods consumed by the individual, business or nation (Chapagain and Hoekstra 2004 ; Chapagain, 2006). However, in Hoekstra and Chapagain (2007) the authors agree that there is a contamination component in the definition of the water footprint. Therefore, in this study it is suggested to complete the definition of the water footprint by including a contamination component. *So, the water footprint (Q_{FP}) will be the total volume of fresh water used to produce goods consumed by the individual, business or nation (consumption component, Q_{FP}*) plus the volume of fresh water needed to somehow assimilate the waste produced by that individual, business or nation (contamination component, Q_{FP}**).* Chapagain and Hoekstra (2004) and Hoekstra and Chapagain (2007) further state that the consumption component of the water footprint, Q_{FP}*, consists of two parts. The first part is the *internal water footprint* (Q_{IFP}). This is the sum of the total annual water volume used from the domestic water resources in the national economy *minus* the annual virtual water flow to other countries related to export of domestically produced products (Q_{VWEdom}). The second part is the *external water footprint* (Q_{EFP}) of a country defined as the annual volume of water resources used in other countries to produce goods and services consumed by the inhabitants of the country concerned (Chapagain and Hoekstra, 2004; Hoekstra and Chapagain, 2007).

2.3 Materials and Methods
2.3.1 Calculation of the water footprint
According to the definition suggested in this paper, the water footprint is

$$Q_{FP} = Q_{FP}* + Q_{FP}** \qquad \dotfill \text{(2.1)}$$

where

Q_{FP}:	The water footprint (m³/year).
$Q_{FP}*$:	The consumption component of the water footprint (m³/year).
$Q_{FP}**$:	The contamination component of the water footprint (m³/year).

As it is difficult to calculate the contamination component of the water footprint, only the consumption component was calculated in this study using equations from (2.2) to (2.6) (Chapagain and Hoekstra, 2004).

$$Q_{FP}* = Q_{IFP} + Q_{EFP} \qquad \dotfill \text{(2.2)}$$

where

Q_{IFP}:	Internal water footprint (m³/year).
Q_{EFP}:	External water footprint (m³/year).

Internal water footprint

$$Q_{IFP} = Q_{AWU} + Q_{IWW} + Q_{DWW} - Q_{VWEdom} \qquad \dotfill \text{(2.3)}$$

where

Q_{AWU}:	The agricultural water use (m³/year).
Q_{IWW}:	The industrial water withdrawal (m³/year)
Q_{DWW}:	The domestic water withdrawal (m³/year).
Q_{VWEdom}:	The virtual water content of exported products produced domestically (m³/year).

In this study, the Q_{DWW} was calculated from PWA's data base (PWA, 2004); it includes the industrial water withdrawal Q_{IWW}. The *agricultural water use* Q_{AWU}, defined as the total volume of water used in the agricultural sector was calculated according to the methodology described in Chapagain and Hoekstra (2004). It includes both effective rainfall the portion of the total precipitation retained by the soil so that it is available for crop production (FAO, 2000) and the part of irrigation water used effectively for crop production.

External water footprint
The Q_{EFP} was calculated according to equation (4) (Chapagain and Hoekstra, 2004).

$$Q_{EFP} = Q_{VWI} - Q_{VWEre-export} \quad \text{.. (2.4)}$$

where

Q_{VWI} :   The virtual water content of imported agricultural and industrial products (m³/year).

$Q_{VWEre-export}$:   The virtual water content of re-exported products (m³/year).

The virtual water of imported crop products has been calculated according to the methodology described in Chapagain and Hoekstra (2004). To calculate the virtual water content of imported industrial products, Q_{VWII}, the net value in US$/year of imports (NVI) was calculated for the years 1998-2002 (Chapagain and Hoekstra, 2004).

$$Q_{VWII} = NVI \times WUV \quad \text{.. (2.5)}$$

where
Q_{VWII} :   Virtual water content of the industrial imports (m³/year).
NVI :   Net value of imports in (US$/year).
WUV :   Global average water withdrawal per unit value of imports (m³/US$).

The per capita consumption component of the water footprint Q_{FPc}* (m³/cap/year) was calculated according to equation (6) (Chapagain and Hoekstra, 2004)

$$Q_{FPc}* = \frac{Q_{FP}*}{Totalpopulation} \quad \text{.. (2.6)}$$

Appendixes 1 and 2 contain example calculation.

2.3.2 Data sources

Raw data about the water quantity from wells and springs and annual rainfall was collected for the period 1988-2003 from the Palestinian Water Authority (PWA, 2004). The Palestinian abstraction was calculated from the PWA's database (PWA, 2004), while the Israeli abstraction was taken from (PWA, 2001). The domestic and agricultural water abstraction from wells and discharge from springs were calculated by taking the sum of the abstraction from all wells and the discharge from all springs for each year and calculating the average abstraction or discharge and the standard deviation thereof for the years 1988 to 2003. Wells with zero abstraction and springs with zero discharge for the last three years were excluded from the calculations in this study. There is a slight decrease in the trend of the rainfall in the West Bank during the period of 1988 to 2003, so the average of the rainfall was used to estimate the total amount of precipitation entering the West Bank.

The FAO food balance sheet for the years 1998-2003 were used as the basis for the food consumption in order to calculate the virtual water in the crops and livestock consumed by Palestinians. The food balance sheet indicates the consumption for the West Bank and Gaza Strip together. To calculate the consumption for the West Bank, all numbers were multiplied by 0.64, the ratio of the population in the West Bank to the total population (West Bank and Gaza Strip) for the years 1998-2003. Data about industrial imports were taken from the PCBS (2004).

2.4 Results and Discussion

2.4.1 Water balance

The West Bank receives 540 mm of precipitation annually (PWA, 2004), this equals a total incoming flow from precipitation (Q_P) of 2970 million m^3/year out of which 679 million m^3/year infiltrates to the groundwater aquifers (Q_I) (Oslo II agreement, 1995). The runoff (Q_R) is about 77 million m^3/year and about 7 million m^3/year are harvested in rain water harvesting systems (Q_{Rh}). Therefore, the total evapotranspiration (Q_{ET}) is 2207 million m^3/year (Figure 2.1).

Abed and Wishahi (1999) indicated that the West Bank receives annually a total quantity of rain between 2,700 – 2,900 million m^3/year. According to the Oslo II agreement (1995) the estimated quantity of water that infiltrates into the groundwater aquifers (Q_I) is 679 million m^3/year (22.9%). Rofe and Raffety (1965) cited in Abed and Wishahi (1999) estimated this quantity as 24.6% for the year 1964/65. In this study the Oslo II agreement (1995) estimates were used to establish the water balance for the West Bank. According to Abed and Wishahi (1999), Rofe and Raffety (1965) estimated the average runoff (Q_R) in the West Bank at 2% of the rainfall while GTZ (1996) estimated it at 3.2%. In this study the runoff flow was taken as 2.6%, the average of the GTZ(1996) and Rofe and Raffety (1965) estimates. Based on this estimation the runoff in the West Bank was found to be about 77 million m^3/year. The population of the West Bank is harvesting (Q_{Rh}) about 7 million m^3/year from rainwater for domestic purposes (MOPIC, 1998 b). Therefore the total evapotranspiration (Q_{ET}) can be estimated to be 2,207 million m^3/year (74.3%). This figure is close to that given by Rofe and Raffety (1965) cited in Abed and Wishahi

(1999), who estimated the evapotranspiration as 69.1% of the total precipitation for the year 1963/64.

Water is abstracted from the groundwater basins by Palestinians and Israelis. Table 2.2 presents the annual Palestinian and Israeli abstraction rates from the three basins through wells and springs. The numbers in Table 2.2 exclude some 170 million m³/year brackish water abstracted or discharged from the aquifers.

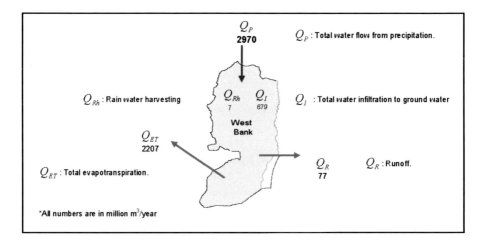

Figure 2.1 Water balance for the West Bank

From this information it can be seen that the total water abstraction (fresh and brackish) by both Palestinians and Israelis amounts to 778 million m³/year while the recharge is only 679 million m³/year which result in an overuse of the ground water.

Table 2.2 Annual recharge and abstraction by Palestinians' and Israelis' from the three basins in the West Bank

Basin	Recharge estimates (million m³/year)			Palestinian abstraction (million m³/year)				Israeli Abstraction* (million m³/year)		
	Ref 1	Ref 2	Ref 3	This study	Ref 1	Ref 2	Ref 4	Ref 1	Ref 2	Ref 5
Eastern	172	172	213	62	54	69	61	40	40	32
Northeastern	145	145	124	31	42	30	31	103	103	99
Western	362	362	376	24	22	22	24	340	344	348
Total	679	683	713	117	118	121	116	483	487	479
Ref 1: Numbers based on Oslo II Agreement (1995)										
Ref 2: Numbers based on Eckstein and Eckstein (2003); Mimi and Aliewi (2005)										
Ref 3: Numbers based on Rofe & Raffety (1963), Ref 4: Numbers based on PWA and USAID (1997)										
Ref 5: Numbers based on SUSMAQ and PWA (2001)										

2.4.2 Water footprint

The consumption component of the water footprint, Q_{FP} *,of the West Bank was found to be 2,791 million m³/year.The internal water footprint, Q_{IFP}, is 1,346 million m³/year and the external water footprint, Q_{EFP}, is 1,445 million m³/year (Figure 2.2).

Consumption component of the water footprint

Internal water footprint

The Palestinians in the West Bank are consuming groundwater for domestic, agricultural and industrial purposes. As can be seen in Figure 3, the total water abstracted from local resources by the Palestinians in the West Bank (Q_{PalAb}) from wells and springs is 117 million m³/year out of which 83 million m³/year is used for agricultural purposes (Q_{AWW}) (irrigating crops and livestock). 34 million m³/year is used for domestic and industrial purposes ($Q_{DWW} + Q_{IWW}$). Moreover, the Palestinians in the West Bank are using some 7 million m³/year rain water harvested in cisterns (Q_{Rh}) for domestic purposes (MOPIC, 1998 b). The Palestinians of the West Bank also produce rain fed crops using the rain water stored in the unsaturated soil. The agricultural water use, Q_{AWU}, was found to be 1,371 million m³/year out of which 66 million m³/year was exported through exporting crops (Figure 2.2 and Table 2.3). The term Q_{AWU} represents part of the evapotranspiration term of the water balance, it includes both effective rainfall (the portion of rainfall which is available for crop production) and the part of irrigation water used effectively for crop production, and it excludes the irrigation losses. The major amount of Q_{IFP} (1,137 million m³/year) is used for producing oil crops and vegetable oils, which is mainly olives and olive oil.

External water footprint

The external water footprint, Q_{EFP}, of the West Bank was found to be 1,445 million m³/year. This figure is the sum of the virtual water imported through the imports of products (crop products, Q_{VWIc}, animal products, Q_{VWIa}, and industrial imports , Q_{VWII},) minus the virtual water exported in exported products (Figure 2.2 and Table 2.3).

The per capita consumption component of the water footprint

The results of the study indicate that the per capita consumption component of the water footprint in the West Bank is 1,116 m³/cap/year. The figure is less than the global average and less than Israeli and Jordanian figures (Chapagain and Hoekstra, 2004) (see Table 2.4). It can be noted that the domestic part of this figure is far less than that of the neighboring countries: only 36% of that of Jordanian and 21% of the Israeli figure.

The contamination component of the water footprint

As was stated before while defining the water footprint, Chapagain and Hoekstra (2004) did not include the volume of water needed to assimilate the waste produced by the individual, business or nation, thus ignoring the second component of the ecological footprint. However, Hoekstra and Chapagain (2007) addressed the effect of pollution on the water footprint and stated that one cubic meter of wastewater should not count for one, because it generally pollutes much more cubic meters of water after disposal, various authors have suggested a factor of ten to fifty they stated.

Nevertheless, societal use of water generates polluted water which itself is not only unfit for direct societal use but which, when discharged in surface water, makes much of the dilution water unfit for use. If so this polluted water is to be considered part of the water footprint.

Here it is suggested to add a second component, other than the consumption component, to the water footprint, which is the volume of fresh water negatively affected by the activities of consumption and use of the individual, business or nation, contamination component.

Table 2.3 The internal agricultural water use of crops and animals and the net virtual water from agricultural and industrial imports.

Group	Internal Agricultural Water Use $(10^6 m^3/year)$	Internal virtual water exported $(10^6 m^3/year)$	Net agriculture water use $(10^6 m^3/year$	Net virtual water imports $(10^6 m^3/year)$
Crops and crops' products				
Cereals	1,11.4	18.9	92.5	986.9
Starchy roots	10.7	0.7	10.0	0.7
Sugar and sweeteners	0.0	0.0	0.0	106.1
Oil crops	557.4	0.0	557.4	14.9
Vegetable oils	579.4	39.3	540.1	92.1
Vegetables	14.8	1.3	13.5	2.4
Fruits	93.6	5.5	88.1	38.9
Stimulants	0.0	0.0	0.0	98.8
Subtotal	**1,367.3**	**65.7**	**1,301.6**	**1,340.8**
Animal products				
Meat	2.2	0.0	2.2	58.4
Milk	0.8	0.0	0.8	16.5
Eggs	0.4	0.0	0.4	1.7
Subtotal	**3.4**	**0.0**	**3.4**	**76.6**
Industrial products				27.6
Total	**1,370.7**	**65.7**	**1,305**	**1,445.0**
Crops included in the calculations were the crops listed in the FAO food balance sheet excluding the items with zero consumption Cereals list: wheat, rice, barley, maize, rye, oats, millet and sorghum, Starchy roots list: cassava, potatoes, sweet potatoes and yams, Sugar and sweeteners: Sugar raw equivalent, honey, Oil crops list: Soya beans, groundnuts, sunflower seed, rape and mustard seed, cottonseed, coconut, sesame seed, palm kernels and olives, Vegetable oils list: Soya bean oil, groundnut oil, sunflower seed oil , rape and mustard oil , cotton seed oil, coconut oil, sesame oil, palm kernels oil, palm oil, olive oil and maize germ oil, Vegetables list: Tomatoes and onions, Fruits list: Oranges, lemons, grapefruit, bananas, apples, pineapples, dates and grapes, Stimulants: Coffee, tea and cocoa beans, Meat: Bovine meat, mutton and goat meat and poultry meat.				

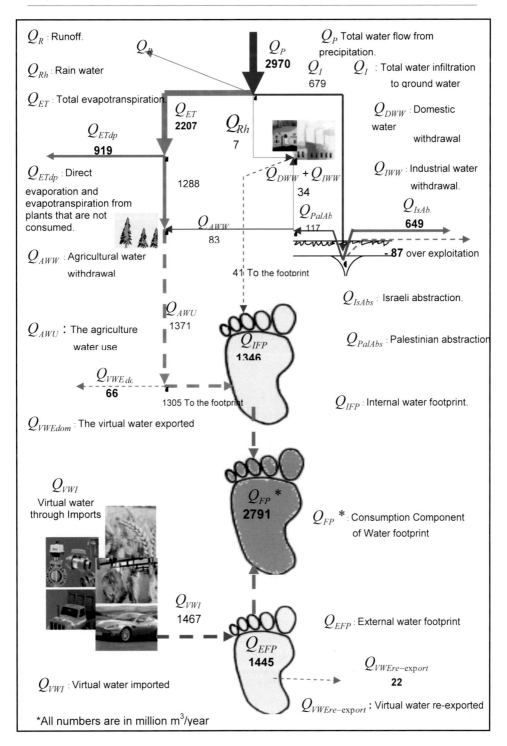

Figure 2.2 Water foot print of the West Bank

Quantifying the second component of the water footprint is a difficult issue. One liter of wastewater has the capacity to contaminate many liters of fresh water if disposed off in a water body without treatment. This is true both for wastewater disposal into surface water as well as through infiltration into the ground water. For example, the WHO limit for lead (Pb) in potable water is 0.01 mg/l. This means that one liter of a wastewater containing 1 mg/l of lead will need 100 liters of fresh water to dilute it to the permissible value, so 1 liter of this wastewater has the potential to contaminate 100 liters of fresh water if disposed in a water body without treatment. Considering the occurrence of self purification, this number may be lower for "clean" organic wastewater. On the other hand the limits for various organic and inorganic constituents of wastewater limits are significantly below that of lead increasing the extent of the contamination component proportionally. In the West Bank, the wastewater in most cases is disposed off into the wadis without treatment. It is difficult to estimate how much fresh water will be contaminated from wastewater infiltrating into the groundwater. This wastewater has the potential to contaminate the shallow aquifers, but deep aquifers may be considered protected from contamination from wastewater infiltration. In any case this means that the contamination component of the water footprint will be "many" times greater than the consumption component making the already scarce resource even more scarce.

Table 2.4 The per capita consumption component of the water foot print of the West Bank and of neighboring countries

Country	Water footprint	Water footprint by consumption category				
		Domestic	Agricultural		Industrial	
	Per capita m³/cap/yr	Internal m³/cap/yr	Internal m³/cap/yr	External m³/cap/yr	Internal m³/cap/yr	External m³/cap/yr
West Bank	1,116	16	548	541	Included in the domestic	11
Jordan	1,303	44	301	908	7	43
Israel	1,391	75	264	694	18	339
Egypt	1,097	66	722	197	101	10
Global average	1,243	57	907	160	79	40
* The figures of the West Bank were calculated in this study while the figures of Jordan, Israel, Egypt and global average were taken from (Chapagain and Hoekstra, 2004 ; Chapagain, 2006; Hoekstra and Chapagain, 2007)						

2.4.3 Water availability, water scarcity and the traditional throw away approach
Total water availability

The water issue in the West Bank is complicated, partly because of the political situation in the area. The aquifers are controlled by Israel. However, Article 40 of the Oslo II Agreement (1995) defines the quantity of water which the Palestinians are allowed to withdraw from their aquifers regardless of how much water is available in these aquifers. So the total water available for the Palestinians in the West Bank was estimated at 198 million m³/year. This number is the sum of the water withdrawal from wells, springs and rainwater harvesting cisterns (123 million m³/year) plus 75 million m³/year agreed upon in the Oslo II Agreement as the future needs of the Palestinians in the West Bank. Therefore, if one assumes that the 2.5 million Palestinians have got the water available for them through Oslo II agreement in 2005, then the totally available water is

80 m³/cap/year in 2005. And if not, the water availability will be the same as the consumption, that is the total per capita water consumption in the West Bank will be 50 m³/cap/ year. In both cases the West Bank can be classified as in the conditions of water scarcity according to Falkenmark's (1986) definition.

According to Falkenmark (1986) "a country whose renewable fresh water availability is less than 1,700 m³/cap/year experiences periodic or regular "water stress". When fresh water availability falls below 1,000 m³/cap/year countries experience chronic "water scarcity". The situation is becoming more severe in the future because of the rapidly growing population from 2.5 million in 2005 to 4.4 million by 2025 (PCBS,1999), which means that in that specific year the per capita water availability will drop to 45 m³ (Figure 2.3).

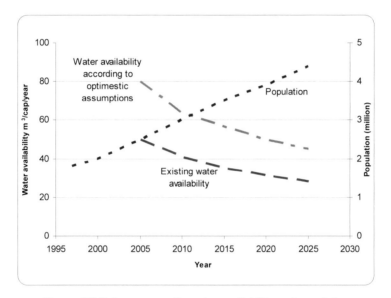

Figure 2.3 Future per capita water availability and population

Domestic water consumption

According to the results of the study the Palestinians in the West Bank are consuming16 m³/cap/year (equal to 44 l/cap/day) for domestic and industrial purposes. The figure is significantly less than the WHO guidelines for the per capita requirement for domestic needs to maintain good health (150 l/cap/day). The figure is also far below the domestic water consumption of the neighboring countries Israel 75 m³/cap/year (205 l/cap/day) and Jordan 44 m³/cap/year (120 l/cap/day) (Table 2.4) (Chapagain and Hoekstra, 2004).

It should be noted that the above concept of water scarcity is determined by assuming that the water is used once before thrown away. Present water management practices will, therefore, increasingly identify conditions of water scarcity because of dwindling resources in combination with increasing population. The common approach of high per capita water consumption, therefore, needs urgent review (Figure 2.4 A) so as to arrive at a situation where the environmental impact of both domestic and industrial water use are significantly reduced (Figure 2.4 B). A large range of options to achieve this significant reduction exist or are in the phase of research testing.

A- common approach of dealing with water

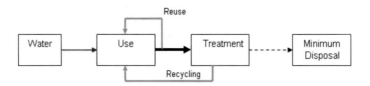

B- Future approach of dealing with water

Figure 2.4 Traditional and future approaches of dealing with water (pls. notice the size of the arrows)

2.5 Conclusions

The objective of the study was to calculate the water footprint for the Palestinians in the West Bank. Within the limitations of the research the following conclusions were drawn:

1. The consumption part of the per capita water footprint ($Q_{FPc}{}^{*}$) in the West Bank was found to be 1,116 m^3/cap/year out of which only 50 m^3/cap/year was withdrawn from local water resources. The contamination component was estimated many times larger than the consumption component making the water footprint many times larger.

2. According to the commonly accepted limits, the West Bank is suffering from a severe water scarcity.

3. The approach of 'use, treat and reuse' may help to improve the situation of water scarcity.

Acknowledgment

The author would like to thank the Palestinian Water Authority PWA in general and especially engineer Adel Yasin for his help in providing data about the water resources in the West Bank.

References

Abed, A. and Wishahi, S. KH. (1999) Geology of Palestine West Bank and Gaza Strip, Palestinian Hydrology Group (PHG), Jerusalem, In Arabic

Allan, J.A. (1997) "Virtual water": A long term solution for water short Middle Eastern economies? Water Issues Group, School of Oriental and African Studies (SOAS), University of London, Paper presented at the 1997 British Association Festival of Science, Water and Development Session, September 9, 1997.

Applied research Institute of Jerusalem ARIJ (1998) Water Resources and Irrigated Agriculture in the West Bank , Applied research Institute of Jerusalem, Palestine

Chapagain, A. K (2006), Globalization of Water, Opportunities and Threats of Virtual Water Trade, PhD Dissertation, UNESCO-IHE institute for Water Education, Delft University of Technology, Delft-The Netherlands.

Chapagain, A. K and Hoekstra, A. Y. (2003), Virtual Water Flows Between Nations in Relation to Trade in Livestock and Livestock Products, Value of Water Research Report series No. 13, UNESCO-IHE institute for Water Education, Delft-The Netherlands. http://www.waterfootprint.org/Reports/Report13.pdf., accessed August 2006.

Chapagain, A. K and Hoekstra, A. Y. a (2004), Water footprint of Nations, Value of Water Research Report series No. 16 volume 1, UNESCO-IHE institute for Water Education, Delft-The Netherlands.
http://www.waterfootprint.org/Reports/Report16.pdf., accessed August 2006.

CIA, (2005) CIA-The world fact book- West Bank.
http://www.cia.gov/cia/publications/factbook/geos/we.html., accessed January 2006.

Eckstein, Y. and G, (2003) Groundwater Resources and International Law in the Middle East Peace Process, Water International, Vol 28, International Water Resources Association, IWRA Executive office , southern Illinois University, Carbondale, U.S.A. pp.154-161.

Falkenmark,M.(1986), Fresh water- time for a modified approach, Ambio, 15(4): 192-200.

FAO (Food and Agriculture Organization of the United Nations), (2000), Crops and Drops: Making the best use of land and water. http://www.fao.org/landandwater/aglw/oldocsw.asp, accessed April 2006.

FAOSTAT data, (2005), Food Balance Sheet for the years 1998-2003. http://apps.fao.org. accessed April 2005.

Germany Technical Cooperation (GTZ) (1996) Middle East Regional Study on Water Supply and Demand Development, Phase 1, Palestine

Hoekstra, A. Y. and Chapagain, A. K (2007), Water footprint of nations: Water use by people as a function of their consumption pattern. Water Resources Management Journal, volume 21, pp 35-48.

Hoekstra, A. Y. and Hung, P.Q (2002), Virtual Water Trade, A Quantification of virtual water flows Between Nations in relation to international Crop trade, Value of water Research report Series No. 11. IHE, Delft, the Netherlands.

Mimi, Z. and Aliewi, A. (2005) Management of Shared Aquifer Systems: A Case Study, the Arabian journal for Science and Engineering, Vol.30, Number 2C. King Fahd

University of petroleum and Minerals, Dhahran, Kingdom of Saudi Arabia. Available online http//:www.kfupm.edu.sa/publications/ajse/, accessed June 2007.

MOPIC (Ministry of Planning and International Cooperation) (1998a) Emergency Natural resources Protection Plan for Palestine "West Bank Governorates", Ministry of Planning and International Cooperation, Palestine.

MOPIC (Ministry of Planning and International Cooperation) (1998 b) Regional Plan for the West Bank Governorates, Water and wastewater Existing Situation, Ministry of Planning and International Cooperation, Palestine.

Oslo II Agreement (1995), Israeli-Palestinian Interim Agreement on the West Bank and the Gaza Strip, Annex III, Article 40, Washington D.C., September 28 1995.

PCBS (Palestinian Central Bureau of Statistics) (1999) Population in the Palestinian Territory 1997-2025 , Palestinian Central Bureau of Statistics, Palestine. http://www.pcbs.org. accessed June 2005.

PCBS (Palestinian Central Bureau of Statistics) (2004) Total Value of Imports and Exports for Remaining West Bank and Gaza Strip by SICT Rev.3 Sections in 1998-2002, Palestinian Central Bureau of Statistics, Palestine, http://www.pcbs.org/trade/tab_04.aspx, accessed June 2005.

PWA (Palestinian water Authority), United States Agency for International Development (USAID) and Camp Dresser and McKee International Inc. (CDM/Morganti) (1997) Task 4 Comprehensive Master Plan for Water Resources in the West Bank and Gaza Strip. Interim report. Water Resources Program. USAID Contract No. 294-0021-C-00-6560-00, Contractor: Camp Dresser and McKee International Inc. Deleverable 4.02.

PWA (Palestinian water Authority) (2004) Data collection by personal communication from the PWA data base, Resources and planning Department, Palestinian Water Authority.

Rofe and Raffety (1963) Geological and hydrological Report, Jerusalem District Water Supply. Central Water Authority, Jordan.

SUSMAQ (Sustainable Management of the West Bank and Gaza Aquifers) and PWA (Palestinian water Authority) (2001) Data Review on the West Bank Aquifers, working report SUSMAQ-MOD #02 V2.0, version2, Water Resources Systems Laboratory, University of Newcstle Upon Tyne and Water Resources and planning Department, Palestinian Water Authority.

WRAP (Water Resources Action Program) (1994) Palestinian Water Resources, A Rapid Interdisciplinary Sector Review and Issues Paper, The Task Force of the Water Resources Action Program, Palestine.

Appendix 1

Calculation of agricultural water use
a. Calculation of agricultural water use for wheat
1. Calculate the Crop Water Requirement for wheat

$$CWR = \sum_{d=1}^{lp} ET_c \ [2].$$

ET_c , crop evapotranspiration (mm) $ET_c = K_c \times ET_0$,

ET_0 , reference evapotranspiration in (mm),

K_c , crop factor,

lp , length of the growing period [days],

ET_0 , K_c and lp were taken from (Chapagain and Hoekstra, 2004).

CWR for wheat = 533 mm

2. Calculate the Specific Water Demand

$$SWD = \frac{CWR}{Yield} \quad \text{(Hoekstra and Hung, 2002: Chapagain and Hoekstra, 2004)}$$

$Yield$ of wheat in the West Bank = 172 kg/1000m^2 (Chapagain and Hoekstra, 2004).

So SWD for wheat = 3098 m^3/ton. Ton = 1000 kg

3. Calculate the agricultural water use

$$Q_{AWU} = SWD * Quantity$$

Quantity refers to the production of wheat in Palestine = 37,000 tons (food balance sheet FAOSTAT data, 2005) as average over 1998-2003.

So Q_{AWU} = 115 X 10^6 m^3.

b. Calculation of total agricultural water use

1. The total internal agricultural water use was calculated by the summation of Q_{AWU} of all crops produced in the area.

2. To calculate the external Q_{AWU}, the imported quantity of each crop was used in the equation in 3 instead of production quantity.

3. The exported Q_{AWU} was then calculated using the exported quantity. Imported and exported quantities taken from (Food balance sheet FAOSTAT data, 2005) as average of 1998-2003.

4. For the crop products such as oil and sugar the Q_{AWU} was multiplied by the value fraction of the product and divided by the product fraction. Value fraction and product fraction were taken from Chapagain and Hoekstra (2004).

Appendix 2

Calculation of virtual water of animal products
Total virtual water for animal

$$VWC_{total} = VWC_{drink} + VWC_{serv} + VWC_{feed} \ (m^3/ton)$$

VWC_{drink} Is the water consumed by the animal for drinking (m^3/1000kg animal)
VWC_{serv} Is the water use for the service of animal such as cleaning (m^3/ton animal)
VWC_{feed} Is the virtual water needed to produce the food for the animal (m^3/ton animal)

VWC_{drink} and VWC_{serv} were taken from Chapagain and Hoekstra (2003)

The VWC_{feed} for animals produced domestically was taken zero because the virtual water for the feed was included in the calculations of the Q_{AWU} for crops which includes the crops consumed by animals.
For imported animal products VWC_{feed} was included and was taken from Chapagain and Hoekstra (2003).

Q_{AWU} for animal products $= \dfrac{VWC_{total} \times valuefraction}{productfraction} \times Quantity$ (produced or imported),

Value fraction and product fraction was taken from Chapagain and Hoekstra (2004).

Ton= 1000kg

Chapter 3

Financial, Environmental and Social Evaluation of Water Management Options in the West Bank, Palestine

Submitted as:
Dima W. Nazer, Maarten A. Siebel, Pieter Van der Zaag , Ziad Mimi and Huub J. Gijzen, Financial, Environmental and Social Evaluation of Domestic Water Management Options in the West Bank, Palestine, Water Resources Management.

Financial, Environmental and Social Evaluation of Domestic Water Management Options in the West Bank, Palestine

Abstract

Water is one of the most valuable natural resources in Palestine. Due to its limited availability, it is a resource that needs thorough protection. Moreover, the current lack of efficient effluent management schemes results in deterioration of the quality of the water resources, thereby further reducing water quantity available for safe consumption. Therefore, it is very important that Palestinians manage their water resources adequately to insure water sufficiency for future generations. Although agriculture consumes most of the water (70%), the domestic sector is not to be neglected. The aim of this study is to evaluate domestic water management options suitable for Palestinian conditions that contribute to achieving water sufficiency in the domestic water use in the *house of tomorrow*. By combining the three pillars of sustainability, the options were evaluated economically, environmentally and socially using the concept of Life Cycle Impact Assessment (LCIA). Results of the study showed that by introducing a combination of domestic water management options, a substantial decrease in the water consumption of more than 50% can be achieved, thereby reducing the pressure on the scarce water resources. The annual environmental impact of the in-house water use can be reduced by a range of 8% when using low-flow shower heads up to 38% when using rainwater harvesting systems. Some of the options (faucet aerators, low-flow shower heads and dual flush toilets were found financially attractive) with a pay back period less than there expected lives, others (rainwater harvesting, gray-water reuse and dry toilets) were found financially unattractive because of the high investment. In the social context, it was found that introducing such options can improve the quality of life of those not having enough water. There is a popular willingness to take part in water conservation in the domestic sector in the West Bank. The strongest driving force for using water conservation measures is the awareness that water is a scarce resource. It was concluded that, theoretically, the *house of tomorrow* can be largely independent in terms of water and sanitation. Education and awareness campaigns in the context of water management with focus on non-traditional options such as dry toilets are key to achieve such a house.

Key words: Environmental, financial, house of tomorrow, social, reuse, water management, West Bank.

3.1 Introduction

The continuing growth in population and development increases the pressure on the water resources all over the world. As a consequence, the number of countries dealing with water scarcity is increasing (Mckinney and Schoch, 1998; Cunningham and Saigo, 1999; Yang *et al.*, 2006). This growing water scarcity can clearly be seen in the Middle Eastern region including the West Bank in Palestine. The per capita water availability in 2005 in the West Bank was 80 m^3/year and this number is expected to decrease to 45 m^3/year by the year 2025 (Nazer *et al.*, 2008). With the formal limits of water scarcity having been defined as 1,000 m^3/capita/year (Falkenmark, 1986), the West Bank is suffering from severe water scarcity.

The above definition of scarcity assumes that the water is used once before being disposed off. This will increasingly result in conditions of water scarcity because of dwindling resources. Therefore, this so-called *use and dispose* approach to water needs urgent reconsideration so as to arrive at a sustainable water use where wastewater is viewed as a resource, ready to be used again, the so-called *use-treat-reuse* approach, one in which both the water use and the environmental impact of domestic water use is significantly reduced. In line with this thinking, in the remainder of this paper wastewater is referred to as "used-water". A large range of options exists to achieve significant reduction in overall water use, making each drop of water more productive (Cunningham and Saigo 1999; Matsui *et al.*, 2001; Rosegrant *et al.*, 2002; Nazer *et al.*, 2008).

The objective of this study is twofold:
1. To evaluate the environmental, financial and social impacts of potential domestic water management options using Life Cycle Impact Assessment (LCIA). These options can be alternatives for water saving or conservation, development of new resources and water reuse.
2. To propose a water use scheme for the *house of tomorrow*, in line with the *use-treat-reuse* approach. In this study the *house of tomorrow* refers to separate houses, housing complexes or buildings.

3.2 Background
3.2.1 Sustainable development in domestic water use
Sustainable development was defined by the United Nations' World Commission on Environment and Development as "the development that meets the needs of the present without compromising the ability of future generations to meet their needs" (WCED, 1987). Sustainability has environmental, financial and social dimensions, that is, for an activity to be sustainable it should be ecologically sound, socially acceptable and economically viable (Hauschild *et al.*, 2005).

Although the domestic and industrial sectors together use far less water than agriculture (30% versus 70%), water demand in the domestic sector is growing rapidly. Water sustainability could be achieved through implementing water saving practices and using newly developed water management approaches, including tapping new water resources (rain water harvesting). The existing ways of dealing with water should be changed from the present inefficient use, exemplified by high rates of water losses, high quality of water used to flush away waste and inefficient water fixtures, to an approach in which the quality and quantity of water is tuned to the specific purpose. Some examples are listed, these examples can be categorized as alternatives of developing new resources, water saving or conservation alternatives, and water reuse.

1. Developing new water resources

- rainwater harvesting systems (RWHS); in these systems rainwater is collected from roofs or rocks and stored in cisterns to be used afterwards thus providing an alternative water resource.

2. Water saving alternatives
   - Faucet aerators (FA) and low-flow shower head (LFSH): these devices restrict the amount of water going through the faucet or shower heads, but add air so the flow of water appears the same (Mayer et al., 2004).
   - Dual flush toilets (DFT): these are toilets that use less water than conventional toilets and have two volumes of flushes (Mayer et al., 2004).
   - Dry toilets (DT): toilets that do not use water or use a small amount of water. Some of these toilets separate urine from faeces (Matsui *et al.*, 2001). These toilets have dual advantage, less water use and less chances of water resources destruction due to pollution. Moreover, these are considered options for resource recovery as the waste can be used for energy and fertilizer production.
   - Leakage reduction (LP): this refers to fixing leaking appliances.

3. Water reuse alternatives
   - Gray-water reuse systems (GWRS): gray-water is defined as untreated, used household water from showers, bathrooms, wash basins and washing machines (Bennett, 1995).

3.2.2 Cleaner production

Cleaner Production is commonly defined as *"the continuous application of an integrated preventive environmental strategy applied to processes, products and services to increase overall efficiency and reduce risks to humans and the environment"* (UNEP, 1999). Although this definition needs a slight adaptation in the context of urban water, the essential elements of Cleaner Production, *i.e.* the proper choice of materials, process efficiency, reuse and recycling of materials and least impact treatment with recovery of resources, remain equally valid (Gijzen, 2001; Siebel and Gijzen, 2002). If so, one can say that Cleaner Production thinking penetrated the way we deal with water: 130 to 500 l/cap/day of drinking water are used, but less than 10 liters are actually needed for drinking and food preparation. The difference is largely used to flush away waste (Gijzen, 1998; Siebel and Gijzen, 2002).

3.2.3 Life cycle impact assessment (LCIA)

Life cycle impact assessment (LCIA) is the process of quantifying the impacts of products, processes and services over the entire period of their life cycle. In carrying out a LCIA, all major impacts should be taken into account (UNEP, 1996; EEA, 1997). LCIA was first developed in the seventies and has been further developed and become the standard when comparing alternatives. More recently a similar approach has been used in the financial area to quantify financial impacts of products, processes and services (Barrios *et al.*, 2008). Moreover, developments in LCIA are moving towards the inclusion of the social dimension (Udo de Haes *et al.*, 2004; Hauschild *et al.*, 2005; Siebel *et al.*, 2007). A combination of the three dimensions of sustainable development (environmental, financial and social) results in an attractive approach, but at the same time poses the risk of impossible data and resource requirements (Udo de Haes *et al.*, 2004).

Environmental life cycle impact assessment (E-LCIA)

The production or use of a product or the rendering of a service involves a number of steps such as extracting and converting resources, manufacturing products, providing infrastructure for transportation and manufacturing, using the product or service, and recycling or disposing off the product that no longer serves its purpose. These steps consume resources, produce pollution and, hence, cause environmental impacts. After an inventory phase, the impacts are categorized and assessed so as to be expressed in a single quantity. In this study, the environmental impact was assessed using the Eco-indicator 99 method (Goedkoop *et al.*, 2000).

Financial life cycle impact assessment (F-LCIA)

In analogy to E-LCIA, F-LCIA is defined as the process of quantifying the financial impacts of a product, process or service during all phases of its life cycle (production, operation and disposal). There are mainly three groups of financial elements that may play a role in F-LCIA:

1. One-time cost   these relate to investment for the acquisition of land area, the construction of buildings, the purchasing of production or transport equipment and, at the end of the lifetime, the costs of disposal of facilities.

2. Recurrent costs   these relate to operational costs such as cost of labor, of raw materials and energy, of operating equipment and of maintenance. In addition, revenues from the sale of products or services are recurrent costs.

3. Cost-related parameters   these relate to interest, inflation and depreciation.

F-LCIA can be used to compare the life cycle costs of system alternatives serving the same purpose but doing so in different ways (Siebel *et al.*, 2007). A cost comparison of alternatives can be done on the basis of a true financial comparison of alternatives taking into account all present and future costs. The present worth (PW) is one such method which relates the cost of any activity at a certain time to the cost at another time given certain values for discount rate (Philippatos and Sihler, 1991; Blank and Tarquin, 2005). The PW of costs and benefits can be calculated according to equation (3.1).

$$PW = A\left[\frac{(1+k)^n - 1}{k(1+k)^n}\right] - I_0 \dots\dots\dots\dots\dots\dots\dots\dots\dots\dots\dots\dots\dots\dots (3.1)$$

Where,

PW :   Present worth, is the monetary value at present or at time zero (US$).

A :   Net annual benefits,(US$/year)

k :   Discount rate,

n :   Number of years (year).

I_0 :   Investment in year zero (US$).

Social life cycle impact assessment (S-LCIA)

In analogy to E-LCIA and F-LCIA, S-LCIA is defined as the process of quantifying the social impacts of a product, process or service over its life cycle. The production of a product, or the rendering of a service or the change therein, almost always have consequences not only in the environmental and financial areas but may also affect those directly influenced by these changes, *i.e.* the consumers/users or those involved in the production of the products or the provisions of the services (Siebel *et al.*, 2007). For example, as a result of a technological innovation, relatively low quality jobs may be replaced by a few relatively high quality jobs causing unemployment for many. It can be argued that a decrease in the availability of water may result in a decrease in the quality of health or may strain social (family) relations. Therefore, the level of happiness of those involved may be affected. On the other hand, water sufficiency will positively impact upon the quality of physical and social health and, therefore, increase the level of happiness.

In conventional development theory, "happiness" equals money and prosperity. It was measured by Gross National Product (GNP). In 1972, Jigme Singye Wangchuch, the newly crowned king of the Himalayan Bhutan Kingdom, created the concept of Gross National Happiness (GNH) to measure prosperity rather than GNP. The king maintains that the economic growth does not necessarily lead to contentment, and instead focuses on the four pillars of GNH: economic self-reliance, a pristine environment, the preservation and promotion of Bhutan's culture and good governance in the form of democracy (Ezechieli, 2003; Bandyopadhyay, 2005).

"Happiness" as a unit of social impact was first used by Hofstetter *et al.* (2006) who listed an extensive number of happiness enhancers and gave each a certain weight. Layard (2005) stated that happiness is a feeling that varies over time throughout our life. The average happiness is determined by one's pattern of activities, one's nature and attitudes and by key features of one's situation, social relationship, health, worries, .etc. There are countless sources of happiness and countless sources of pain and misery. Layard (2005) stated that there are seven enhancers of happiness *i.e.* family relationships, financial situation, work, community and friends, health, personal freedom and personal values. Similarly, Siebel *et al.* (2007) proposed five categories of happiness, job quality, quality of physical health, quality of social health, earthly possessions and various.

Siebel *et al.* (2007) developed a simple approach to quantify the social impact of societal or industrial activities. The approach is based on the assumption that each person starts out with a given value of happiness, expressed as Socio-points in S-LCIA (in analogy with Eco-points for E-LCIA and Euro-points for F-LCIA). Over the years, this initial happiness value will change as a result of concrete physical or emotional experiences such as a painful accident, not passing an important exam, a prison sentence, or from less identifiable influences such as having been born into a happy family, having had a bad youth or having a born physical handicap. Some of these experiences will reduce the initial value of happiness, others will increase it. By categorizing the various social impacts into the various happiness categories, and by equating each social impact with the maximum value in that category, the happiness of people can be quantified or the change in happiness of people can be assessed when determining the happiness value before and after the change. By determining the happiness values various times before and after a specific social change, the short and long-term impact of a certain social change can be determined.

3.2.4 Statistical methods

Chi-square test: can be used for testing the change in a specific characteristic between two groups due to some kind of treatment for one of the groups. This can be done by calculating the observed value of the Chi-square and comparing it with the critical value of Chi-square under the required level of significance (Chase and Bown, 1986).

3.3 Materials and Methods
3.3.1 Research approach

An overview was made of potential domestic water management options on the basis of their financial, environmental and social impacts. One options focus on developing new water resource, *i.e.* rainwater harvesting system (RWHS) and one focus on water reuse, *i.e.* grey-water reuse system (GWRS). Moreover, four indoor water saving options were evaluated; two were related to the toilet as largest indoor water consumer in the West Bank (34%), *i.e.* the dual flush toilet (DFT) and the dry toilets (DT). Also the low flow shower head (LFSH) was chosen to cover the bath and shower water consumption (22%) and the faucet aerators (FA) was chosen to cover the bathroom sink and the kitchen with 14% and 13% respectively of the indoor water use. Remaining consumption (17%) (laundry, cooking and drinking and house cleaning) was relatively small (Nazer *et al.*, 2007).

The comparison of the options was made on the basis of evaluating the expected change (increase or decrease) in the environmental, financial and social impacts due to the implementation of these options. The do-nothing alternative was the reference for calculating the change. An inventory analysis was carried out to determine the changes associated with the implementation of each option, *i.e.* water use, used-water production and energy consumption. The calculations were based on the per capita water consumption in the West Bank distributed over the different indoor water consumption points (toilet, kitchen, bath, laundry...etc (Nazer *et al.*, 2007).

3.3.2 Life cycle impact assessment

LCIA was chosen to evaluate the change (increase or decrease) in the environmental, financial and social impacts as a result of the implementation of the studied options. Although LCIA involves the three phases of the life cycle, *i.e.* production (manufacturing /construction and installation), operation and disposal/demolition, the analysis in this study is made for the production and operation phases because systematic investigation about disposal and demolition of house-hold sanitary systems does not exist in the West Bank.

Environmental life cycle impact assessment (E-LCIA)

The following steps were carried out:

1. The environmental impact of producing (manufacturing /constructing and installing) each option was determined using the Eco-indicator 99 database (Goedkoop et al., 2000) in which the environmental impact was given in eco-point/unit weight of material. For each option, the materials used in the production of that options were investigated in terms of type and weight, and then the environmental impact was calculated.

2. On the basis of the inventory analysis of implementing the water management options, the changes in pollutant emissions to air, water and soil resulting from the changes in water use, used-water production and energy consumption were determined. According to these emissions the change in environmental impact was determined.

Financial life cycle impact assessment

In order to evaluate the financial costs of each water management option the following steps were carried out:

1. Determination of the investment in (US$).
2. Determination of the annual operational costs (US$/year).
3. Determination of the annual savings in operational costs (US$/year).
4. Determination of the present worth of the costs and benefits (US$/year).

Calculations were carried out for a period of 10 years. Discount rate was estimated at 5% per year (CIA, 2009). This rate is the average discount rate of Israel and Jordan provided by CIA fact-book (2009) because a discount rate for Palestine is not available. Operation and maintenance were estimated at 5% (Hutton and Bartram, 2008).

Social life cycle impact assessment

To approach the social aspects of the various water management options in this study, the impact of these options on the level of happiness of people using or proposed to use these options was assessed. A questionnaire was prepared for this purpose; the questionnaire contained both closed and open questions. The closed (yes/no) questions were about the willingness to use and to pay for the different options, while the open questions were about the reasons behind accepting or rejecting certain options. After an explanation of each option, the respondents were asked if they were willing to use an option and the reasons behind willing to use or not willing to use it. The respondents were also asked if they were willing to pay for the option in question and why. The respondents could give one or more reasons for each question.

Three groups with a total of 244 adults were involved in the study. The first group (92 participants) was chosen from general users of water from different segments of society: students, employees, house-wives. These people were approached through workshops in different localities in the West Bank. The workshops (Figure 3.1) began with an introduction about the West Bank water resources and the water scarcity in the area.

Thereafter, the in-house water consumption pattern was explained followed by an overview of potential options available for reducing in-house water use and their associated financial and environmental costs and benefits. Then the participants were asked to fill out the questionnaire. The second group (120 participants) filled out the questionnaire with help of university students who first explained the options. The third group (32 participants) was formed by professionals in the water sector.

Figure 3.1 Workshop of filling out the questionnaire

The average initial value of happiness of a Palestinian in the West Bank was set at 500 Socio (happiness)-points. This initial value is expected to change as a result of life experiences. In this study the difference between living in a country characterized by a situation of serious and increasingly serious water stress *vs.* living in a country in which this situation of water stress is reduced as a result of measures meant to reduce water consumption and, therefore, water stress, was quantified and related to the happiness of the people. The questionnaire was meant to quantify the change in the feelings of people relative to the situation before and after taking appropriate measures to improve water availability (*i.e.* reduce water stress). In doing so, it was assumed that if a person is willing to use an option, then using this option increases the initial value of happiness of this person by a value of a socio-points because of the perception of contributing to a good cause, a better life, .etc. Furthermore, if that person is willing to pay for that specific option, the value of happiness of the person increases further by another value of b socio-points. On the other hand, if a person is rejecting (equal to: not willing to use) the option, the happiness level of this person is expected to drop by a socio-points as a result of using this option. The use of a particular option could, for example, become enforced as part of a government measures to reduce water stress. If this person is not willing to pay for this specific option, it means that paying for that option is expected to further decrease the level of his happiness by b socio-points.

In a group of people, the overall level of happiness is influenced by the level of variation in happiness of this group. The happiness level may increase for some participants, may decrease for others, or remain the same. Accordingly, the change in happiness of a group of people is the increase in happiness of some minus the decrease in happiness of others. Therefore, the change of happiness for a group of people, or the social impact as it will be called in the rest of the paper, can be calculated according to equation 3.2.

$$\Delta S = a\alpha_u + b\beta_p - a\alpha_{nu} - b\beta_{np} \dots\dots\dots\dots\dots\dots\dots\dots\dots (3.2)$$

where

ΔS :   Change in the level of happiness or the social impact of a group (socio-points),

α_u :   Willingness to use, given as the fraction of respondents willing to use the option (range:0-1),

β_p :   Willingness to pay, given as the fraction of respondents willing to pay for the option (range:0-1),

α_{nu} :   The fraction of respondents not willing to use the option (range:0-1),

β_{np} :   The fraction of respondents not willing to pay for the option (range:0-1),

a , b :   Constants describing the effect of not having water on the person's level of happiness.

Given that respondents were asked to express either their willingness or not willingness to use or pay for (yes/no answers), this implies that:

$\alpha_{nu} = 1 - \alpha_u$ and $\beta_{np} = 1 - \beta_p$.   Therefore equation (3.2) becomes

$$\Delta S = 2a\alpha_u + 2b\beta_p - (a+b) \dots\dots\dots\dots\dots\dots\dots\dots\dots (3.3)$$

In case of a situation of serious water stress, it is assumed that willingness to pay approaches the willingness to use a certain option. In that case a and b are equal and the equation (3.3) simplifies to

$$\Delta S = 2a(\alpha_u + \beta_p - 1) \dots\dots\dots\dots\dots\dots\dots\dots\dots (3.4)$$

In order to determine the values of a and b , a group of 52 people, randomly chosen from the telephone book of the residents of West Bank, were asked how much their happiness would be affected if they wanted to do something that needed water such as taking a shower, washing, cleaning, ...etc. only to find out that there would be no water to do so. They were asked to rate the ups or downs of their happiness level on a scale of 0-100.

The response of the respondents suggested that the level of happiness will decrease by 70%. The situation of not having water when needed may affect two of the big seven enhancers of happiness proposed by Layard (2005), *i.e.* health and family relationships.

Assuming that each of the big seven enhancers has an equal effect on happiness, then not having water will affect $2/7^{th}$ of the total decrease in happiness (70%). Therefore, the initial level of happiness will be decreased by (2/7 of 70%) = 20%. So a value of 100 (20% of 500 the initial average) socio-points for the total decrease or increase in the level of happiness (a plus b) could be reasonable. Assuming that a and b are equal, then each of them will take a value of 50 socio-points.

3.3.3 List of assumptions regarding water management options

There are a variety of ways to produce the chosen water management option which vary significantly with regard to the materials used for producing or constructing them, cost, environmental impact …etc. For example, a rainwater harvesting reservoir could be made of plastic, concrete or any other material according to which the cost and the environmental impact may vary significantly. Therefore, assumptions were made in this study for the sake of the calculations:

1. The average family size in the West Bank is 6 (PCBS, 2007).
2. The water consumption pattern is according to Nazer et al., (2007).
3. It is assumed that dry toilet do not use any, are made of polypropylene and have a total weight of 13 kg. Calculation of the compost for dry toilet is based on a production of 12 kg/person/ year (Matsui et al., 2001). Price of compost was estimated on the basis of local market at $US 0.3/kg of compost (personal communication).
4. A dual flush toilet uses one full flush/cap/day (6liters) and 2 reduced flushes/cap/day (3liters) (EPA, 1995). It is made of ceramics with a total weight of 35 kg.
5. A gray-water reuse system uses the water used in the sink, bath, shower and laundry. It consists of a plastic reservoir and plastic piping system.
6. A Low-flow shower head saves about 40% of the water used by traditional shower heads (EPA, 1995). They are made of plastic with a weight of 0.15 kg.
7. Faucet aerators can reduce the water used in the faucets by 60% (EPA, 1995). They are made of plastics.
8. Leakage in the West Bank accounts for some 30%-40% (PECDAR, 2001). Due to system improvements and increased awareness of water scarcity, leakage prevention saves 10% of the total consumption.
9. Rainwater harvesting, these are reinforced concrete reservoirs with an average volume of the cistern of 80 m^3 (MOPIC, 1998). The size of the cistern was cross-checked using the methodology developed by Van der Zaag (2000); the idea is to calculate the size of a rain water harvesting storage tank that can satisfy a household for a given satisfactory level (30-90%). Knowing the daily water demand for the household, the catchment area (usually a roof) and the daily rainfall in the region for some 10 years, the size of the storage tank can be calculated (Van der Zaag, 2000).

3.4 Results and Discussion

3.4.1 Inventory of change in resource use

Table 3.1 presents the annual resource (water and energy) use and the used-water (wastewater) production before implementing (do-nothing alternative) and after implementing the different water management options together with the change in resource use.

3.4.2 Environmental Impact

Table 3.2 presents the environmental impact of the production (manufacturing/ construction) and the annual reduction in the environmental impact due to the implementation of the different water management options as well as the annual net reduction in the environmental impact. Although the environmental impact of constructing the rain water harvesting system is very high relative to other options, it is the most environmentally sound option with respect to the net reduction in the environmental impact followed by the dry toilet and gray-water reuse systems. In the case of the dry toilet, the reduction in the environmental impact, due to the use of compost produced in the toilet, as a substitute fertilizer, was not included in the calculation. Nevertheless, this will increase the environmental benefits. It can be seen that the largest fraction of environmental impact reduction is from energy reduction.

3.4.3 Financial impact

Initial investments for the management options vary significantly between options according to type, manufacturer and place of installation. For example, costs of gray-water systems vary from simple inexpensive diverters which cost less than $200 a piece to complex treatment, storage and irrigation systems that cost several thousands of dollars (PLANETARK, 2007). Cost calculations in this study were based on investment costs given by the specified reference in Table 3.3.

From Table 3.3 it can be seen that the present worth (PW) of the rainwater harvesting gray-water reuse, and dry toilets is negative which means that these are financially unattractive. This can be explained by the high investment needed for these alternatives. However the social benefits gained from improved health and the gain in productive time resulted from improved health, time saving associated with better access to water and sanitation and economic gains associated with saved lives were not included in the calculations; Hutton *et al.* (2007) estimated the average rate of return of these benefits at a global average of US$ 8.1 per dollar invested for combined water supply and sanitation. Moreover, the calculations in this study were based on the existing prices of water and energy which are expected to rise in the future due to the increasing water scarcity which would increase the benefits of any water saving system.

The faucet aerator was found the most financially feasible on annual basis followed by, low flow-toilet and low-flow shower heads (Table 3.3). The pay back period for faucet aerators, low-flow shower heads, dual flush toilets options, was less than their expected life. In this context Mayer *et al.* (2003, 2004) said that an analysis of cost and benefit for installing low flow-toilets and low-flow shower heads showed that these products pay for them selves in water and sewer cost savings. The payback time for installing low-flow toilets (from savings on water and wastewater charges) was under 2 year, for a showerhead was 1.6 years and for a faucet aerator was under 1 year.

3.4.4 Social impact

Figure 3.2 presents the expected change in the level of happiness of people as a result of implementing the different options for the three groups of people investigated. Table 3.4 presents the detailed results of the group who did not attend the workshops as an example. It can be seen that the most socially acceptable water management options are leakage prevention (LP) and rainwater harvesting (RWHS). The dry toilet (DT) is the least socially acceptable.

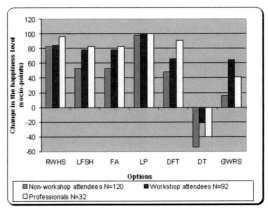

Figure 3.2 Expected change in the level of happiness as a result of implementing water management options among the sampled societal groups.

Results of the study showed that attending the workshops succeeded in improving the participants' awareness about water conservation options, including such non-traditional options as the dry toilets. Results indicated that there is a significant increase in social impact of some water management options on the participants of the different groups investigated. The social impact of using dry toilets was minus 54 socio-points in the non-workshop attendees (participants who did not attend workshops); the impact increased to minus 21 socio-points among the workshop attendees (participants who attended the workshops). Social impacts of gray-water reuse, dual flush toilets and faucet aerators also increased among the workshop attendees relative to the non-workshop attendees (Figure 3.2).

To statistically explain the relationship between the level of awareness and the willingness to use and pay for each option, or in other words to explain the difference in the change of happiness between the workshop attendees and the non-workshop attendees, the chi-square test was used.

Table 3.1 Annual reduction per family (6 persons) in resource use due to the implementation of domestic water management options in the operational phase

Option type	Option		New resource	Water saving options					Reuse option
		Do-nothing alternative	Rainwater harvesting system	Low-flow shower head	Faucet aerator	Leakage Prevention	Dual flush toilet	Dry toilet	Gray-water reuse system
Resource use after using the option	Groundwater use (m^3/year)	83	3	76	69	75	70	55	47
	UW production [c] (m^3/year)	83	83	76	69	75	70	55	47
	Energy consumption [b] (KWh/year)	324	68	296	269	293	273	215	183
Reduction in resource use due to using the option	Groundwater use (m^3/year)	0	80	7	14	8	13	28	36
	UW production [c] (m^3/year)	0	0	7	14	8	13	28	36
	Energy consumption [b] (KWh/year)	0	256	28	55	32	51	109	141

[a] All calculations were made for a family of six and on the basis of the water consumption pattern given by (Nazer et al., 2007).

[b] Calculations of energy use based on the following information:

1. The energy use for water abstraction and transportation is 3.2 KWh/m^3 (JWU, 2007).
2. No energy needed for water treatment because the only treatment used is chlorination (JWU, 2007).
3. The energy use for wastewater treatment is 0.7KWh /m^3 (Al_Bireh, 2006).

[c] UW stands for used-water (wastewater)

Table 3.2 Environmental impact of production, annual reduction and net reduction in the environmental impact for the different water management options relative to the "do-nothing" alternative.

Option type		New resource	Water saving options					Reuse option
Option		Rainwater harvesting system	Low-flow shower head	Faucet aerator	Leakage prevention	Dual flush toilet	Dry toilet	Gray-water reuse system
Environmental impact due to production [a]								
Total impact of production (mPt)		167000	148	1.5	-	980	364	8880
Expected life (year)		50	10	10	-	20	20	10
Annual environmental impact of production (mPt/year)[b]		**3344**	**15**	**0.2**	**-**	**49**	**18**	**888**
Reduction of environmental impact due to implementation								
Water treatment	Water use reduction (m³/year)	0.0	7	14	8	13	28	36
	Chlorine reduction (kg/year)[c]	0.0	0.0	0.01	0.0	0.01	0.02	0.02
	Envi. Impact reduction (mPt/year)[c]	**0.0**	**0.0**	**0.38**	**0.0**	**0.38**	**0.76**	**0.76**
Used-water (waste-water) production	Reduction (m³/year)	0.0	7	14	8	13	28	36
	COD reduction (Kg/year)[d]	0.0	12	22	13	21	45	58
	Envi. impact. reduction (mPt/year)[e,f]	**0.0**	**35**	**63**	**37**	**60**	**130**	**167**
Energy	Reduction (kWh)[g]	256	28	55	32	51	109	141
	Envi. Impact reduction (mPt/year)	**6656**	**728**	**1430**	**823**	**1326**	**2834**	**3666**
Total reduction of environmental impact (mPt/year)		**6656**	**763**	**1493**	**860**	**1386**	**2965**	**3834**
Net reduction of environmental impact (mPt/year) = total reduction – annual impact of production		**3312**	**748**	**1493**	**860**	**1337**	**2947**	**2946**
Percentage reduction compared to the do-nothing (%)*		**38**	**8**	**17**	**10**	**15**	**33**	**33**

*Please note that the environmental impact of the do-nothing alternative is 8806 (mPt/year)

All calculations were made for a household of a family of six persons

[a] The environmental impact of production was calculated using the Eco-indicator 99 (Goedkoop et al., 2000).

[b] The impact has been annualized, converted to reflect the impact on yearly basis.

[c] Water treatment: only chlorination is used, 0.6 mg/l is used (JWU, 2007). Environmental impact= 38 mPt/kg chlorine (Goedkoop et al., 2000).

[d] COD concentration = 1586mg/l (Mahmoud et al., 2003).

[e] mPt is the unit of environmental impact in milli eco-points

[f] Environmental impact of 1 kg COD is 2.88 mPt (Nazer et al., 2006) and for kWh energy is 26 mPt (Goedkoop et al., 2000).

[g] The energy use for water abstraction and transportation is 3.2 kWh/m³, no energy needed for water treatment because the only treatment used is chlorination (JWU, 2007). The energy use for wastewater treatment is 0.7kWh /m³ (AI Bireh, 2006

Table 3.3 Annual benefits, costs and the present worth of net benefits for the different water management options relative to the "do-nothing" alternative.

Option Type	New resource	Water saving options					Reuse option
Option	Rainwater harvesting system	Low-flow shower head	Faucet aerator	Leakage Prevention	Dual flush toilet	Dry toilet	Gray-water reuse system
Investment							
Initial Investment (cost and installation)	4000	15	45	10/year	200	900	700
Expected life of equipment	50	10	10	-	20	20	10
Investment for 10 years (I_0)	**800**	**15**	**45**	**100**	**100**	**450**	**700**
Reference number	3	1	1	5	1	4	2
Annual operational benefits							
From water savings ($)	96	8	17	10	16	34	43
From wastewater treatment savings ($)	0	2	4	2	4	8	11
From energy savings ($)	36	4	7	4	7	15	20
From using compost as fertilizer ($)	na	na	na	na	na	22	na
Total annual benefits ($)	132	14	28	16	27	79	74
Annual operational costs							
Operation and maintenance cost ($)	-40	-1	-2	0	-5	-23	-35
Net annual benefits ($)	**92**	**13**	**26**	**16**	**22**	**56**	**39**
Total benefits –Total costs							
Present worth PW *	**-92**	**85**	**155**	**23**	**69**	**-19**	**-400**
Pay back period (year)	<1	<1	<1	<1	1.6		

* PW is calculated according to the equation 3.1 (Blank and Tarquin, 2005) ; k is the real discount rate = 5 %
*** Average cost of water for domestic use = 1.2 US$/m^3 (PWA, 2007); Average cost of energy = 0.14 US$/KWh (Al-Bireh, 2006); Cost of wastewater treatment =0.3 US $ /m^3 (Al_Bireh, 2006); Calculations were made over a period of 10 years; Annual operation and maintenance 3% of investment; Cost of compost is estimated on the basis of local market at $US 0.3/kg of compost;a person produce some 12kg/year (Matsui et al., 2001)
na stands for not applicable or not available
References for the investment: 1: (Mayer et al., 2003), 2: (PLANETARK, 2007), 3: (Krishna, 2005), 4: (UNESCO-IHE, 2007)
5: it was assumed that US$ 10/year will be spent for leakage fixing.

Table 3.4 Expected change in the level of happiness as a result of implementing water management options for the group who did not attend the workshops

Option Type		New resource	Water saving options					Reuse option
Option		Rainwater harvesting system	Low-flow shower head	Faucet aerator	Leakage Prevention	Dual flush toilet	Dry toilet	Gray-water reuse system
Willingness to use	Total number of respondents	120	119	119	120	121	120	120
	Number of respondents willing to use the option	118	108	108	120	100	37	82
	Number of respondents not willing to use the option	2	14	14	0	21	83	38
	α_u : the fraction of respondents willing to use the option	0.98	0.9	0.9	1	0.83	0.31	0.68
Willingness to pay	Total number of respondents	120	119	119	120	121	120	120
	Number of respondents willing to pay for the option	102	75	75	118	78	18	58
	Number of respondents not willing to pay for the option	18	44	44	2	42	102	62
	β_p : the fraction of respondents willing to pay for the option	0.85	0.65	0.65	0.98	0.65	0.15	0.48
Social impact ΔS *		83	53	53	98	48	-54	16
Willingness to pay / willingness to use $\dfrac{\beta_p}{\alpha_u}$		0.87	0.72	0.72	0.98	0.78	0.48	0.71

* Social impact (change in happiness ±) was calculated according to the equation 3.4 $\Delta S = 2a(\alpha_u + \beta_p - 1)$

** (+) means increase in happiness and (−) means decrease in happiness

The calculated chi-square value for dry toilets, grey-water reuse systems and faucet aerators exceeded the critical value 3.84 at 5% level of significance (Table 3.5) meaning that attending the workshops had a significant positive effect on the participants willingness to use and pay for the presented options. However, for rainwater harvesting systems and leakage prevention there was no significant difference between the two groups (Table 3.5). This can be explained by the fact that these options are well-known and widely used in the West Bank. There was also a significant difference between the professionals and the group of the non-workshop attendees. The social impact was greater in the professionals group. This can be explained by the fact that the professionals are well aware of these water management options because of their professional background in the water sector. However, the social impact of dry toilets and grey-water systems in the professionals group is less than the impact in workshop attendees group. This may be because the professionals are educated with the idea of " water is plentiful available" and "all problems of water can easily be solved" (just like a medical doctor is raised with the concept of "treatment first"). Therefore, in the professionals mind, tools that save water or eliminate the use of water in clear cases of potential negative impact on the public health, are simply not done, these are outside their professional interest.

It can be concluded that making people aware of certain water management options, including such non-traditional options as the dry toilets, increases the willingness to use and pay for these options and, hence, the social impact (change in happiness level) due to using these options.

Table 3.5 The calculated and critical value of Chi- square between the workshop attendees group and the non-workshop attendees group for the different water management options

Option		Rainwater harvesting system	Low-flow shower head	Faucet aerator	Leakage Prevention	Dual flush toilet	Dry toilet	Gray-water reuse system
Calculated chi-square	Willingness to use	0.7	3.3	3.3	0.0	3.6	6.2	17.8
	Willingness to pay	0.04	8.2	8.2	1.3	2.1	5.3	10.1
Critical value of chi-square is 3.84 at 5% level of significance, a value < 3.84 means no significant difference between the groups and a value >3.84 means there is a significant difference(Chase and Bown, 1986).								

It was found that the willingness to pay for an option is less than the willingness to use that option, $\alpha_u \leq \beta_p$ and therefore $\dfrac{\beta_p}{\alpha_u} \leq 1$ (Table 3.4). However, the willingness to pay for the option is expected to increase with increasing importance of that option. For example, a person may be willing to use a rainwater harvesting system but may not be willing to pay for it if the availability of water is not a critical issue. But if water availability is a daily returning issue, that person's willingness to pay will approach his or her willingness to use. From Figure 3.3, it can be seen that $\dfrac{\beta_p}{\alpha_u}$ for options considered important such as leakage prevention is approaching 1 while that for options considered not important it is less than 1. Therefore, efforts should be devoted towards increasing the awareness about the options which are considered of low importance.

Reasons behind accepting or rejecting water conservation options
The respondents of the three groups were asked about the reasons behind their willingness or un-willingness to use and to pay for the options. Detailed information about the reasons can be found in Appendix 1.

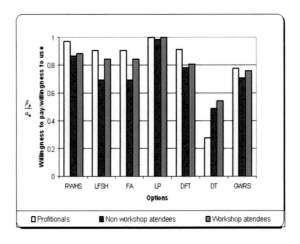

Figure 3.3 Willingness to pay/willingness to use, $\dfrac{\beta_p}{\alpha_u}$ for the different water management options

Dry toilets (DT): Participants who rejected DT explained their rejection for emotional, religious, public health and financial reasons amongst others. The emotional reason was the dominant: 48% of participants who reject these toilets said they just could not use these toilets. Public health concerns scored 30%, religious concerns scored 18% of those not willing to use dry toilets (122). Those who said they were willing to use these toilets (90 respondents) did so for reasons of water saving (63%), money saving (9%), environmental saving (10%).

Gray-water reuse systems (GWRS): Participants who are willing to use GWRS (160 respondents) did so for reasons of water saving (49%), money saving (6%) and environmental saving (4%). Public health concern was the most dominant reason for rejecting these systems (42%) followed by the emotional reasons (22%) of those not willing to use GWRS (46 respondents).

Dual flush toilets (DFT): 75% of the participants who are not willing to use DFT (28 respondents) often felt that these devices are water wasting because of the large volume of water these devices use relative to the volume of hand flushing containers which they use. It is common in the area to use special containers to flush away the waste. The volume of these containers varies between 1-2 liters.

Rainwater harvesting systems (RWHS): Almost all workshop participants (98%) were willing to use RWHS. These are considered as an alternative water resource to ensure water security when the piped water supply is cut off. Many participants suffer from loosing access to water especially in summer when water is cut off for long periods. Some participants (5%) said that it is the only water resource they have. 86% of the workshop attendees and 85% of the non-workshop attendees have the willingness to pay for constructing a RWHS regardless of the high cost of these systems.

3.4.5 Relating environmental, financial and social impacts

Figure 3.4 shows a strong correlation between environmental and financial benefits of the options during the operational phase: the coefficient of determination R (0.89) is greater than the critical value of R (0.708) for a level of significance of (0.01) (Chase and Bown, 1986). For one eco-point decrease in the environmental impact, there is an increase in profit of US$ 107.

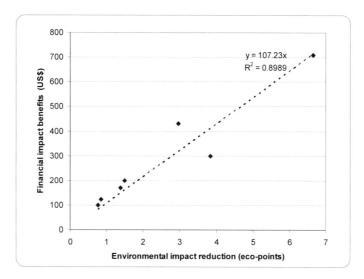

Figure 3.4 Relationship between environmental impact and financial impact of the operational phase

In contrast, it was found that the relationship between the social impact and both financial and environmental impacts is weak, the coefficients of determination R, 0.52 and 0.45 respectively, are less than the critical value of R (0.708) for a level of significance of 0.01 (Chase and Bown, 1986). This can be explained by the reasons behind the participants' willingness to use and pay for the options: the most motivating reason was water availability (46%) rather than money saving (6%) or environmental concerns (4%). Moreover, the Palestinians have long ago adapted to water scarcity as a fact of life. This can be well illustrated by the fact that only 5% of the water use is withdrawn from existing groundwater resources and 43% is consumed from rain water through consuming rain-fed crops. In contrast, 52% of that water is imported through imported goods (Nazer *et al.*, 2008).

3.4.6 The *house of tomorrow*

According to Nazer *et al.* (2007) the average in-house water consumption in the West Bank is 38.1 $\pm$ 18.4 liter/cap/day (Table 3.6). The large variation means that the in-house water consumption varies considerably from place to place. Nevertheless, the numbers in Table 6 were used to show to what extent, even in conditions of limited water use, further water saving is feasible by combining different water management options in the so-called *house of tomorrow*. It can be seen that due to the implementation of these options a significant reduction can be achieved in the per capita in-house water consumption from 38.1 liter/cap/day to a minimum of 15.7 liter/cap/day (Table 3.6). The combination proposed in this context is: dry toilet instead of water-based toilet, low-flow shower heads and faucet aerators, potentially resulting in over 50% saving in water (Nazer *et al.*, 2007). A gray-water reuse system can be installed through which the used-water can be reused for irrigating the garden, by doing so the outdoor consumption can be eliminated. According to the numbers presented in Table 3.6, a family of six will only need 34m^3/year which could easily be harvested from rain water from a roof of 100 m^2 area.

In the proposed *house of tomorrow* a combination of six options was used. However, in some places some of these options may not be applicable. In such cases other combinations of two or three options could be used according to the specific conditions. For example in places where a dry toilet is not applicable, a low-flow toilet may be a substitution.

In the above analyses, dry toilets play an important role, in spite of the fact that these presently, among those questioned, are the least socially acceptable water conserving device of those investigated. However, it should be noticed that water availability predictions for the West Bank are rather dire. Less and less water will be available through the years to come leaving no other choice than to go for the most effective water conservations options. Timely planning and awareness raising are, therefore, vital ingredients in averting what is bound to come if no serious measures are taken in time (Nazer *et al.*, 2008).

Pilot projects can play a significant role in awareness raising. Areas without water supply or sanitation system and very low water consumption can be fertile grounds for implementing pilot projects. It would be an opportunity for these areas to improve the water and sanitation systems. In this context Matsui *et al.* (2001) stated that the existing sanitation systems in the developed countries are not enough to protect the environment, since environmental protection was not an issue when these systems were developed.

Therefore, developing countries, like Palestine, have an opportunity to skip the phase of water wasting sanitary facilities and to move directly into the implementation of water conscious sanitary equipment.

The introduction of these water management options should, in the very near future, be oriented towards a situation in which each single- and multi-family housing property is largely independent in terms of water and sanitation systems. Each property has its own water resource (RWHS) and its own used-water treatment system (gray-water treatment system in combination with a dry toilet). This can be achieved by adapting building codes accordingly. However, this may not be possible in some places where the rainwater harvesting does not yield sufficient water. In such places water can be supplied from other resources which will be easy due to the reduced demand in other areas. At the same time these areas can still reduce the consumption by using other water management options.

Table 3.6 House-hold water consumption on a per capita basis without implementing water management options nd expected per capita water consumption with implementation (Nazer *et al.*, 2007).

Point of use		Per capita water consumption without		Per capita water consumption with	
		Average (l/c/d)	Standard deviation	Proposed water management option	(l/c/d)
Bathroom	Toilet flushing	12.9	9.5	Dry toilet instead of water based toilet	0.0
	Sink	5.3	4.1	Faucet aerators	2.1
	Bath and shower	8.3	7.1	Low-flow shower head,	5.0
Kitchen	Dish washing	5.0	3.5	Faucet aerators	2.0
	Cooking and drinking	1.6	1.5	-	1.6
Cleaning	Laundry	3.0	3.1		3.0
	House cleaning	2.0	2.3	-	2.0
Total in-house		38.1	18.4		15.7
Outdoor *		6.7		Gray-water use	0.0
Total		44.8			15.7
*outdoor water use was estimated at 15% of the total house use according to (PWA,2005) Rain water reservoir with a minimum capacity of 40 m³ could be installed to guarantee the provision of the water needs of a family of 6 and a gray-water reuse system can also be installed which can make use of the 15.7 liters of the used-water in the house.					

Figure 3.5 presents the existing *use and dispose-* based water use system. Figure 3.6 presents the proposed *use-treat-reuse-* based system which allows for closing the cycle of water use. In-between, a variety of choices exists to reduce domestic water use in existing buildings.

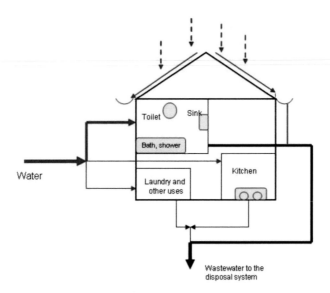

Figure 3.5 The existing situation of water use according to the *"use and dispose"* approach.

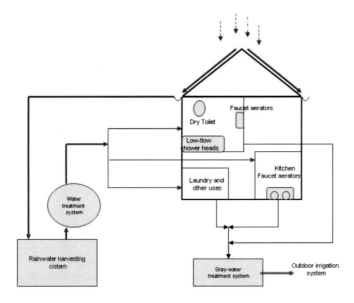

Figure 3.6 The proposed water use in the *house of tomorrow* according to the *"use-treat-reuse"* approach.

3.5 Conclusions

Within the limitations of the study the following conclusions were drawn:

1. Although the average per capita daily water consumption is very low in the West Bank, significant water savings can be achieved by using water conservation options. Introducing the appropriate options can reduce the in-house water use by more than 50%.

2. By implementing water management options the annual environmental impact of the in-house water use can be reduced by a range of 8% for low-flow shower heads up to 38% for rainwater harvesting systems relative to the do-nothing alternative.

3. Some of the options investigated were found financially attractive such as faucet aerators, low-flow shower heads and dual flush toilets. However, others (rainwater harvesting, gray-water reuse systems and dry toilets) were found unattractive because of the high investment needed. However, the social and health benefits may justify these investments.

4. There is a popular willingness to take part in water conservation in the domestic sector in the West Bank. The strongest driving force for using water conservation measures is the awareness that water is a scarce resource.

5. Water awareness is significantly affecting willingness to invest in water saving devices. Education and awareness campaigns in the context of water management may need to focus on non-traditional options such as dry toilets.

Acknowledgement

Thanks are extended to all organizations and individuals (employees, students, general people...etc) who have participated in the workshops or helped in filling out the questionnaires, out of which the data presented in this study were obtained. The participation of school students in measuring the indoor water consumption is highly appreciated. Thanks are due to all schools (staff and students) who participated in this study. A special thank you to Sima Kuhail for her dedicated help in implementing the workshops related to the domestic water use and processing the data collected from these workshops, Layana Nazer and Lina Afifi for their help in collecting data.

References

Al-Bireh Municipality (2006) Personal communications with engineer Naief Tumaleh, resident engineer in Al-Bireh Wastewater Treatment Plant. Al-Bireh, Palestine.

Barrios, R., Siebel, M., Helm, A., Bosklopper, K. and Gijzen, H. (2008) Environmental and Financial Life Cycle Impact Assessment of Two Plants of Amsterdam Water Supply, Journal of Cleaner Production, volume 16(4) pp 471-476.

Bandyopadhyay, S. (2005) Gross National Happiness and Beyond: A micro welfare economics approach, paper presented in the Second International Conference on Gross National Happiness, St. Francis Xavier University, Antigonish, Nova Scotia, Canada, June 2005. available online http://www.gpiatlantic.org/conference/papers , accessed November 2008.

Bennet, D. (1995) Graywater, An Option for Household Water Reuse, Home energy Magazine online. Retrieved January 2004 http://homeenergy.org/archive.

Blank, L. and Tarquin, A. (2005) Engineering Economy, 6th ed, International edition, McGrawHill.

Chase, W., and Bown, F. (1986) General statistics, John Wiley and Sons, Inc. New York.

Cunningham, W. P. and Saigo, B. W. (1999) Environmental Science a global concern,5th ed. Mc graw-Hill Companies, United States of America.

European Environmental Agency (EEA) (1997), Life Cycle Assessment, A guide to Approaches, Experiences and Information Sources, available on line, accessed October 2007.
http://reports.eea.europa.eu/GH-07-97-595-EN-C/en/Issue%20report%20No%206.pdf

Environmental Protection Agency (EPA) (1995) Cleaner Water through Conservation, U.S. Environmental Protection Agency, U.S.A. http://www.epa.gov/waterhome. accessed February 2004.
Ezechieli, E. (2003) Beyond Sustainable Development: Education for Gross National Happiness in Bhutan, Monograph of the degree of Master of Arts, Stanford University. Available on line accessed November 2008.
http://suse-ice.stanford.edu/monograghs/Ezechieli.pdf ,

Falkenmark,M.(1986), Fresh water- time for a modified approach, Ambio, 15(4): 192-200.

Gijzen H.J. (1998) Sustainable wastewater management via re-use – Turning Waste into Wealth. In: Garcia, M., Gijzen, H.J., Galvis, G. (eds.), Proc. AGUA_ water and Sustainability, June 1-3, 1998, Cali, Colombia, pp 211-225.

Gijzen, H. J. (2001) Aerobes, anaerobes and phototrophs – a winning team for wastewater management. Published in *Wat Sci Tech*. Vol 44, # 8, pp123-132.

Goedkoop, M., Effting, S. and Collignon, M. (2000), The Eco-indicator 99, A Damage Oriented Method for Life Cycle Impact Assessment, Manual for Designers, second edition, Pre consultants, Amersfroot, available on www.pre.nl. Accessed June 2008.

Hauschild, M., Jeswiet, J. and Alting, L. (2005) From Life Cycle Assessment to Sustainable Production: Status and Perspectives, Annals CIRP, 54(2), pp 535-555.

Hofstetter, P., Madjar, M. & Ozawa, T. (2006) Happiness and Sustainable Consumption: Psychological and Physical Rebound Effects at Work in a Tool for Sustainable Design, Int. J LCA 11, Special Issue. 1, pp 105-115.

Hutton, G., Haller, L. and Bartram, J.(2007), Economic and health effects of increasing coverage of low cost household drinking-water supply and sanitation interventions to countries off-track to meet MDG target 10, Public Health and the Environment, World Health Organization (WHO), Geneva

Hutton, G. and Bartram, J., (2008), Global costs of attaining the Millennium Development Goal for water supply and sanitation Bulletin of the World Health Organization, 86 (1), pp 13-19.

Jerusalem Water Undertaking (JWU) (2007) Water prices, Rates and Tariffs, Ramallah, Palestine, accessed June 2007 http://www.jwu.org.

Krishna,H.J. (2005) The Texas Manual on Rainwater Harvesting, 3rd edition, Texas Water Development Board, Austen, Texas.

http://www.twdb.state.Tx.us/publications/reports/RainwaterHarvestingManual-3rdedition.pdf. Accessed June 2007

Layard, R. (2005), Happiness, Lessons from a new science, Penguin Books, London, England.

Matsui, S., Henze, M., Ho, G. and Otterpohl, R. (2001) Emerging Paradigms in Water Supply and Sanitation, In Frontiers in Urban Water Management: Dead Lock or Hope, IWA publishing and UNESCO, UK.

Mayer, P. W., DeOreo, W.B, Towler, E. and Lewis, D.M. (2003), Residential Indoor water Consumption Study: Evaluation of High Efficiency Indoor Plumbing Fixture Retrofits in Single-family Homes in the East Bay Municipal Utility District Service Area, prepared for East Bay Municipal Utility District and the United States Environmental Protection Agency (EPA), Aquacraft, Inc. Water Engineering and management, Boulder, Colorado. Available on line: www.aquacraft.com. Accessed July 2007.

Mayer, P. W., DeOreo, W.B, Towler, E., Martien, L. and Lewis, D.M. (2004), Tampa Water Department Residential Water Consumption Study: The impacts of High Efficiency Plumbing Fixture Retrofits in Single-family Homes, prepared for Tampa Water Department and the United States Environmental Protection Agency (EPA), Aquacraft, Inc. Water Engineering and management, Boulder, Colorado. Available on line: www.aquacraft.com. Accessed July 2007.

Mckinney,M.L. and Schoch, R.M (1998) Environmental Science, Systems and Solutions, Web enhanced edition, Jones and Bartlet publishers INC, Sudbury, Massachusttes, USA.

Ministry of Planning and International Cooperation (MOPIC) (1998) Regional Plan for the West Bank Governorates, Water and Wastewater Existing Situation 1st ed., Ministry of Planning and International Cooperation, Palestine.

Nazer, D.W., Al-Sa'ed, R.M. and Siebel, M.A. (2006), Reducing the environmental impact of the unhairing-liming process in the leather tanning industry, Journal of Cleaner Production vol.14 # 1:pp 65-74.

Nazer, D.W., Siebel, M.A., Mimi, Z. and Van der Zaag, P. (2007), Reducing Domestic Water Consumption as a Tool to Raise Water Awareness in the West Bank, Palestine, paper presented in the 13th International Sustainable Development Research Conference, Vasteras, Sweden June 10-12 2007.

Nazer, D.W., Siebel, M.A., Mimi, Z., Van der Zaag, P. and Gijzen, H.J. (2008), Water footprint of the Palestinians in the west Bank, Palestine, Journal of American Water Resources Association (JAWRA), volume 44, issue 2, pp 449-458.

PCBS (Palestinian Central Bureau of Statistics) (1999) Population in the Palestinian Territory 1997-2025 , Palestinian Central Bureau of Statistics, Palestine.

PCBS (Palestinian Central Bureau of Statistics), 2007. The Palestinian Expenditure and Consumption Survey 2006, Levels of Living in the Palestinian Territory, The Final Report (January 2006-January 2007), Palestinian Central Bureau of Statistics, Palestine.

PCBS (Palestinian Central Bureau of Statistics) (2008), Population, Housing and Establishment Census 2007, Census Final Results in the West Bank, Summary (Population and Housing), Ramallah- Palestine.

PECDAR (2001) Palestinian Water Strategic Planning Study, Palestinian Economic Council for Development and Reconstruction.

Philippatos, G. C., and Sihler, W.W. (1991) Financial Management text and cases, 2nd ed., Allyn and Bacon, Massachusetts.

PLANET ARK, (2007), Graywater :the Basics. Graywater fact sheet series, available online at   http://www.planetark.com/products/PAGreyWaterInfo.pdf, accessed June 2008.

Palestinian Water Authority (PWA) (2005) Personal communication, unpublished data, Ramallah, Palestine.

Rosegrant, M. W, Cai, X. and Cline, S. A. (2002) Global Water Outlook to 2025, Averting an Impeding Crisis, Food Policy Report, A 2020 Vision for Food, Agriculture, and the environment Initiative, International Food Policy Research Institute (IFPRI) and International Water Management Institute (IWMI), Colombo, Sirilanka.

Siebel, M.A. and Gijzen, H.J. (2002) Application of cleaner production concepts in urban water management in: Workshops Alcue – Sustainable development and urbanization: from knowledge to action, Science & Technology, Vol.3.

Siebel, M.A., Chang, CT., Gijzen, H.J. and Rotter, V.S. (2007) Life Cycle Impact Assessment- A Case Study of solid waste Collection, Proceedings of the Eleventh international Waste Management and Landfill Symposium, S. Margherita di Pula, Cagliari, Italy; 1-5 October by CISA, Environmental Sanitary Engineering Centre, Italy.

Udo de Haes, H., Heijungs, R., Suh, S., Huppes, G. (2004), Three Strategies to Overcome the limitations of Life-cycle Assessment, journal of Industrial Ecology, Volume 8, # 3, pp19-32.

UNEP (United Nations Environment Program) (1996) Life cycle assessment: what it is and how to do it?, First Edition, United Nations publications. Paris, France.

UNEP (United Nations Environment Programme) (1999) International Cleaner Production Information

Van Der Zaag, P. (2000), Estimating storage requirement for rainwater harvested from roofs, a paper presented in the 4th Biennial Congress of the African Division of the International Association of Hydraulic Research, Windhoek, Namibia.

UNESCO-IHE (2007) personal communication

WRAP (Water Resources Action Program) (1994) Palestinian Water Resources, A Rapid Interdisciplinary Sector Review and Issues Paper, the Task Force of the Water Resources Action Program, Palestine.

WCED (World Commission on Environment and Development) (1987) Our Common Future, Oxford university press, United Kingdom.

Yang, H., Wang, L., Abbaspour, K. And Zehnder, A. (2006) Virtual water and the need for greater attention to rain-fed agriculture, Water 21, Magazine of the International Water Association, April 2006.

Chapter 4

Optimizing Irrigation Water Use in the West Bank, Palestine

Accepted as

Nazer, D.W., Tilmant, A., Mimi,Z., Siebel, M.A., Van der Zaag, P. and Gijzen, H., Optimizing irrigation water use in the West Bank, Palestine. Agricultural Water Management.

Optimizing Irrigation Water Use in the West Bank, Palestine

Abstract

The West Bank is suffering from severe water scarcity: the resources are limited and the demand is increasing. Agriculture is the biggest water consumer as it accounts for 70% of the total water consumption. Therefore, effective agricultural water management is key to achieving water sustainability in the area. This study aims at optimizing irrigation water allocation using a linear mathematical programming model. The analysis was applied in five agricultural zones in the West Bank and included four main vegetable crops namely tomato, cucumber, eggplant, squash and citrus. Three scenarios were analyzed in the study: The first scenario presents the existing cropping patterns, the second maximizes profit under water and land availability constraints and the third one maximizes profit under constraints of water and land availability and local crops consumption. Results of the study showed that changing the cropping patterns of the five crops included in the study under land and water availability constraints reduces the water used for irrigation by 10%, which is equivalent to a reduction of 4% of the total agricultural water use. This would increase the total added value from these crops by 38%, equivalent to 4% of the entire agricultural sector. With regard to scenario 3 the reduction of water use for irrigation was also 10% and the increase in added value was 12% for the studied crops and 1% for the entire agricultural sector. It was concluded that water scarcity can be approached by changing the cropping patterns according to their water use.

Key words: Agriculture, cropping patterns, linear programming, profit maximization, water scarcity.

4.1 Introduction and background
4.1.1 General

With continuous population and economic growth, water resources have become increasingly scarce in a growing number of countries and regions of the world. As the largest water user, food production is directly constrained by water scarcity (Yang et al., 2006). Several authors (Rosegrant et al. 2002; Playan and Mateos, 2006; Yang et al., 2006; Falkenmark, 2007) in discussing the capacity of the earth to produce food for its increasing population, argued that one of the main factors limiting further expansion of food production will be water. This scarce resource is facing heavy and unsustainable demand plus pollution which decreases water quality and, therefore, availability.

The West Bank, as many areas of the Middle East, is suffering from a severe water scarcity. The water use of Palestinians in the West Bank is 50 m^3/cap/year, withdrawn from water resources available in the area. This water is used for domestic, industrial and agricultural purposes. 70% of the water is used for agriculture. Therefore, the West Bank is badly in need of improved water management in the agricultural sector in order to obtain the optimal benefits from the scarce water resources in the area (Nazer et al., 2008).

Water may be a scarce resource but there are many ways of using it more efficiently, that is, making each drop of water more productive (Rosegrant *et al.*, 2002). Falkenmark (2007) suggested three options to capture the additional water needed to meet the requirements of future food production. These are: 1) increasing water productivity by reducing losses, 2) better use of infield rain and expanding rain-fed agriculture, and 3) virtual water options as advocated by Allan, 1997; WWC, 2004; Hoekstra and Hung, 2005; Hoekstra and Chapagain, 2007.

Mathematical programming models are recommended to optimize allocation of scarce resources. These models provide an optimal way of combining scarce resources to satisfy the proposed needs, they analyze situations where the available resources must be combined in a way to optimize an objective function, while meeting various constraints. Mathematical programming is well suited for the analysis of water use in the agricultural activities. These models can provide information on optimal cropping patterns in areas with scarce water, that is, a trade-off between crops and farming practices in the agricultural sector. Results of optimization can be used by agricultural planners and farmers to evaluate their cropping patterns (Loucks *et al.*, 1981; Haouari and Azaiez, 2001; Mimi, 2001; Hillier and Lieberman, 2005).

Several models have been developed world wide in the context of irrigation water management. The aims of these models are different, some seek to maximize profit by irrigation scheduling and allocation of water among competing crops in a fixed area (Vedula and Mujumdar, 1992; Vedula and Kumar, 1996; Sunantara and Ramirez, 1997), others maximize profit by changing cropping patterns using the area as the decision variable in order to allocate water between competing crops (Loucks *et al.*, 1981; Al-Weshah, 2000; Haouari and Azaiez, 2001). No study related to optimization of crop patterns using mathematical programming was undertaken for the West Bank. Therefore, the objective of this study is to find the optimal cropping pattern in the West Bank in order to reduce water use for irrigation while at the same time maximizing profits using a linear programming model.

4.1.2 Available water resources and water use

The West Bank receives on average 540 mm/year of precipitation, this equals a total incoming flow from precipitation of 2970 million m^3/year out of which 77 million m^3/year flows as runoff, 7 million m^3/year is harvested in rain water harvesting systems, the total evapotranspiration is 2207 million m^3/year and 679 million m^3/year infiltrates into the groundwater aquifers (Nazer *et al.*, 2008).

The water issue in the West Bank is complicated, partly because of the political situation in the area. The aquifers are controlled by Israel. In 2005 the water consumption was 50 m^3/cap/year (Nazer *et al.*, 2008). The optimistic scenario about future water availability estimated the per capita available water at 80 m^3/cap/year. This estimation was made on the basis of Palestinians being provided by the existing water withdrawal (123 million m^3/year) plus an extra 75 million m^3/year agreed upon in the Oslo II agreement (1995). The West Bank can be classified as an extremely water scarce area using Falkenmark (1986) definition of water scarcity (1000 m^3/cap/year). The situation will become even worse in the future because of the rapidly growing population from 2.5 million in 2005 to 4.4 million by 2025 (PCBS, 1999), which means that the per capita water availability might drop to 45 m^3/cap/year (Nazer et al., 2008).

Irrigation is currently using 83 million m^3/year equal to 70% of the water withdrawn from groundwater. In addition, water stored in the soil is used to cultivate rain-fed crops such as olives, grapes and many others. This water can not be reallocated for other uses of water unless changes in land use have been made.

4.1.3 Existing situation of the agricultural sector

Table 4.1 presents the total cultivated area in the West Bank as well as the crop (fruit trees, field crops and vegetables) distribution. Each area has two types of irrigation systems, rain-fed and irrigated. Most of the fruit trees are rain-fed, with the area cultivated by olives being the most dominant with 83%, followed by grapes with 6.6%. Some fruit trees are irrigated such as citrus (1% of the total area of fruit trees). Field crops are mostly rain-fed. The most dominant crop is wheat (36%) followed by barley (23%) of the area cultivated by field crops. 68% of the area cultivated by vegetables is irrigated (PCBS, 2006). The main vegetable crops are tomatoes, cucumber, eggplant and squash. These crops account for 51% of the area cultivated by vegetables (Table 4.2). The total annual vegetable production in the West Bank is 340,000 metric tons. The main crops named earlier account for 78% of total vegetable production (PCBS, 2006) (Table 4.2). In this study the focus will be on these main vegetable crops and citrus. These crops consume 36% of the total agriculture water use in the area. Green houses are sometimes used to cultivate tomatoes and cucumber. The area of these green houses was assumed fixed in this study.

Table 4.1 Rain-fed and irrigated area of fruit trees, field crops and vegetables in the West Bank (PCBS, 2006)

	Fruit trees	Field crops	Vegetables	Total
Rain-fed area (hectare)	106,900	43,400	4,000	154,300
Irrigated area (hectare)	2,100	1,500	8,600	12,200
Total	109,000	44,900	12,600	166,500

Table 4.2 Area cultivated by tomatoes, cucumber, eggplant ant squash vegetable crops and the production from each crop

Crop	Area cultivated		Production	
	Quantity (hectare)	Percentage (%)	Quantity (ton)	Percentage (%)
Tomatoes	1,700	13	101,000	30
Cucumber	1,800	14	89,000	26
Eggplant	900	7	43,000	13
Squash	2,200	17	31,000	9
Others	6,000	49	76,000	22
Total	12,600	100	340,000	1000

4.1.4 Linear programming

A linear programming model is applicable for the solution of problems in which the objective function, Z which describes the quantity to be maximized or minimized, and the constraints which indicate the scarceness of the resource, appear as linear functions of the decision variables, x, (Loucks *et al.*, 1981; Walsh, 1985; 2001; Karamouz *et al.*, 2003).

A general linear programming problem is to maximize or minimize a linear function Z:

$$Z = f(x) = c_1 x_1 + c_2 x_2 + ...c_n x_n$$

Subject to the constraints

$$
\left.
\begin{array}{l}
a_{11} x_1 + a_{12} x_2 + ...a_{1n} x_n \leq,=,\geq b_1 \\[2mm]
a_{21} x_1 + a_{22} x_2 + ...a_{2n} x_n \leq,=,\geq b_2 \\[2mm]
.... \\[2mm]
a_{m1} x_1 + a_{m2} x_2 + ...a_{mn} x_n \leq,\geq,= b_m
\end{array}
\right\}
$$

And the non-negativity constraint

$$x_1 \geq 0........x_n \geq 0$$

where
x_i are the decision variables.
a_{ij} and b_i are constants.

In vector-matrix notation the general problem could be written as follows

$$Z = CX \quad\text{...(4.1)}$$

Subject to constraints

$$AX \leq,=,\geq B \quad\text{...(4.2)}$$

$$X \geq 0 \quad\text{..(4.3)}$$

where
C is an *n*-component row vector,
X is an *n*-component column vector,
A is an *m* x *n* matrix, and
B is an *m*-component column vector.

4.1.5 Solving the model

The model has been solved using Microsoft Excel solver. Excel solver provides the optimal solution to the problem (4.1) to (4.3), i.e. the optimal value of the objective function , z^* , the optimal values of the decision variables, x^* and the shadow prices of constraints (4.2) and (4.3). It also provides the sensitivity report which contains information about the decision variables and the constraints. It is important to perform a sensitivity analysis to investigate the effect of change in the parameter values on the optimal solution and in order to identify the sensitive parameters, the parameters whose values cannot be changed without changing the optimal solution.

Shadow prices, which are also available at the optimal solution, indicate how much profit is achieved by changing that constraint right-hand side, B , by one unit. A positive shadow price means an increase of profit equal to the shadow price will be achieved if the constraint was relaxed by one unit. A zero shadow price means that the profit is not affected by increasing or decreasing the constraint; the constraint is nonbinding. A negative shadow price means that by relaxing the constraint by one unit, a decrease in profit equal to the shadow price will be noticed. The allowable range of increase and decrease of the constraint within which the current optimal solution remains feasible, is another piece of information provided by the sensitivity report (Hillier and Lieberman, 2005).

4.1.6 Water productivity

Water value as defined by Rogers *et al.* (1998, 2002) consists of two parts, the economic value and the intrinsic value. The economic value consists of: 1. value to users, 2. net benefits from return flows, 3. net benefits from indirect uses and 4. the adjustment for social objectives. In this study the calculations of the water productivity is limited to the value of water to users.

The average water productivity in the agricultural sector can be expressed as amount of agricultural product by weight per unit water. The production is usually expressed in unit weight (kg or ton) of crop. However, expressing the production in monetary units ($) is more convenient when different crops are to be considered together (Playan and Mateos, 2006). The overall productivity of water for z zones and x crops can be calculated according to equation 4.4

$$WP = \sum_{j=1}^{j=z} \sum_{i=1}^{i=x} \frac{PV_{ij}}{WU_{ij}} \quad\text{...(4.4)}$$

Where

WP is the overall water productivity in ($/m^3$)

PV_{ij} is the value of product i in zone j ($)

WU_{ij} is the water use of crop i in zone j (m^3)

The marginal water value for crop irrigation indicates the value of the last unit of water used. This concept has important implications, in times of water shortages, neoclassical economic theory advises supplying the last unit of water to its most productive uses and thereby maximizing the productivity of available water.

4.2 The linear programming model

A linear programming model was formulated to find the optimal crop patterns in the West Bank in order to reduce water use for irrigation while maximizing net benefit from irrigation. Therefore, increasing the water productivity per unit water. The objective function of the model is to maximize total profit under the constraints of land availability, water availability and production demand. The areas cultivated by each crop present the decision variables.

Maximize total profit, TP,

$$TP = \sum_{j=1}^{j=z}\sum_{i=1}^{i=x} P_{ij} * A_{ij} * Y_{ij} - \sum_{j=1}^{j=z}\sum_{i=1}^{i=x} Cc_{ij} * A_{ij} - \sum_{j=1}^{j=z}\sum_{i=1}^{i=x} Wc_j * A_{ij} * Wd_{ij} \qquad (4.5)$$

where

TP	The total profit achieved from cultivating X crops in Z zones (US$)
P_{ij}	The farm gate price of crop i in zone j ($/1000 kg)
A_{ij}	The area cultivated by crop i in zone j (hectare), decision variable.
Y_{ij}	The yield of crop i in zone j (kg/hectare)
Cc_{ij}	Variable cultivation cost of crop i in zone j ($/hectare)
Wc_j	Cost of water in zone j ($/m^3)
Wd_{ij}	Water required to produce crop i in zone j (m^3/hectare)

Subject to the constraints

a. Land constraint

$$\sum_{j=1}^{j=z}\sum_{i=1}^{i=x} A_{ij} \leq A_a \qquad \dots\dots\dots\dots\dots\dots\dots\dots\dots\dots\dots\dots\dots\dots(4.6)$$

b. Water constraint

$$\sum_{j=1}^{j=z}\sum_{i=1}^{i=x} Wd_{ij} * A_{ij} \leq W_a \qquad \dots\dots\dots\dots\dots\dots\dots\dots\dots\dots\dots\dots\dots(4.7)$$

c. Local consumption constraint

$$\sum_{j=1}^{j=z}\sum_{i=1}^{i=x} A_{ij}Y_{ij} \geq TD \dotfill (4.8)$$

d. Green houses area constraint

$$\sum_{j=1}^{j=z}\sum_{i=1}^{i=x} A_{ijgh} = A_{agh} \dotfill (4.9)$$

e. Non negativity constraint

$$\sum_{j=1}^{j=z}\sum_{i=1}^{i=x} A_{ij} \geq 0 \qquad\qquad (4.10)$$

where

A_a	Total available area for agriculture in all zones (hectare)
W_a	Total water allocated for agriculture (m^3)
TD	Total demand for the agricultural crops (ton=1000 kg)
A_{agh}	Total available area for agriculture in green houses in all zones (hectare)
A_{ijgh}	The area cultivated in green houses by crop i in zone j (hectare)

The model was applied to five different agricultural zones in the West Bank, namely Jenin, Jericho, Nablus, Tulkarem and Tubas. These zones produce 66% of the vegetables in the West Bank and 83% of the citrus. It included the main vegetable crops and citrus. Three scenarios were analyzed, scenario 1 describing the existing cropping patterns in the area and was used as a reference to compare the water use and profit for the other two scenarios. The objective function of scenario 2 was to maximize profit under water and land constraints. For scenario 3, the objective function was to maximize profit under the water, land and local consumption constraints. A constraint was put for scenarios 2 and 3 that restricts the area cultivated by tomatoes and cucumber in green houses to avoid allocation of tomatoes and cucumber for this irrigation regime because it is more profitable compared to open irrigation and because areas occupied by green houses are known and cannot be expanded.

Data collection
A questionnaire was prepared and designed to cover the main irrigated crops in the West Bank. It was distributed to 250 farmers in the agricultural zones of the West Bank. A team of people specialized in completing questionnaires was selected in order to distribute and to help farmers in completing the questionnaire. For practical reasons the members of the

team were selected from the different areas. As a first step, the questionnaire was tested by distributing 40 copies to farmers in the different agricultural zones, this also served as training for the questionnaire fillers. Afterwards the questionnaire was improved in line with the comments from the farmers and the questionnaire fillers. The questionnaires were completed by direct interviews with farmers; the farmers were accessed by the team members in their farms. There is no exact information about the number of farmers in each zone. So it was decided to distribute 50 questionnaires in each zone. The information collected from the questionnaire was about:

1. the variable cultivation cost, Cc_{ij} ,

2. water cost, Wc_j ,

3. farm gate price of crops, P_{ij} ,

4. the water demand, Wd_{ij} , required to produce the different crops in the different zones. Wd_{ij} was based on how much water the farmers apply to the crops.

Information about the total available area, A_a , for agriculture in the different zones and the areas already cultivated by the different crops, A_{ij} existing, and the yield, Y_{ij} , were based on PCBS (2006). The yield was assumed fixed and equal to the maximum because the studied crops belong to the crops with high sensitivity to water deficit, yield response factor (K_y) of these crops is greater than one. According to Doorenbos and Kassam (1979) K_y for citrus is 1.1-1.3 and for tomatoes 1.05. Because K_y value for squash, eggplant and cucumber could not be found in the literature, it was assumed that $K_y > 1$ which is the value given by Doorenbos and Kassam (1979) for a group of crops that is most similar to these crops. Depending on the yield response factor, in conditions of limited water availability, one can maintain yield and reduce the area, or reduce yield while maintaining the same area. When $K_y > 1$ it is more convenient to maintain the yield and reduce the area (see Box 1). The water allocated for the different zones, W_a , was taken from a Palestinian Water Authority report (PWA, 2004).

Total variable cost, $Cost_{To}$, was calculated according to equation 4.11

$$Cost_{To} = Cc_{ij} + Wc_j * Wd_{ij} \quad \text{...(4.11)}$$

The profit achieved from crop i in zone j ($/hectare),   , was calculated according to equation 4.12

$$Pr_{ij} = P_{ij} * Y_{ij} - Cost_{To} \quad \text{...(4.12)}$$

The total crop demand, TD , was calculated by multiplying the yearly per capita crop demand (PCBS, 2007) by the total population.

Box 1: Yield response factor K_y

Yield response factor is a term used to quantify the response of yield to water supply. It relates relative yield decrease
(1-Y_a/Y_m) to relative evapotranspiration deficit (1- Et_a/Et_c) according to equation 13 (Doorenbos and Kassam, 1979).

$$1 - \frac{Y_a}{Y_m} = K_y \left[1 - \frac{ET_a}{ET_c} \right] \qquad \ldots\ldots\ldots\ldots(13)$$

where

Y_a :   is actual yield of the crop [kg ha^{-1}]

Y_m :   is maximum (expected) yield in absence of environmental or water stresses

K_y :   yield response factor

ET_c :   potential (maximum) crop evapotranspiration in the absence of environmental or water

             stresses ($K_c ET_o$)

ET_a :   actual (adjusted) crop evapotranspiration as a result of water stresses.

Application of the yield response factor for planning, design, and operation of irrigation projects allows quantification of water supply and water use in terms of crop yield and total production for the project area. Under conditions of limited water supply the yield decrease for crops with K_y >1 will be greater than the loss of yield for crops with K_y <1. When maximum production for the project area is being aimed at, and the land is not a restricting factor, the available water supply would be towards fully meeting the water requirements for crops with K_y >1.

When K_y >1, for maximum production, the irrigated area is based on the available water supply meeting full crop water requirement $ET_a = ET_c$ and $Y_a = Y_m$ over an area irrigated with crop water requirements fully met. When K_y <1, for maximum production, the irrigated area is based on available water supply partially meeting the crop water requirement $ET_a < ET_c$ and $Y_a < Y_m$ but increased area is maintained (Doorenbos and Kassam, 1979).

4.3 Results
4.3.1 Water use, productivity and profit

Table 4.3 presents the area, water use, total profit and the water productivity for the three scenarios. Under scenario 2 a decrease of 10% in the water use could be achieved combined with an increase of 38% in the added value of the studied crops. A water decrease of 10% was also achieved in scenario 3 combined with an increase of 12% of the added value. The productivity of water was increased by 9% (from 3.2U S$/m^3 to 3.5 US$/m^3) in scenario 2 and 6% (from 3.2 US$/m^3 to 3.4 US$/m^3) in scenario 3.

Table 4.3 Water use, added value and the water value given by the three scenarios studied

Scenario*		Scenario 1	Scenario 2	Scenario 3
Area (hectare)		5330	5300	5310
Water use ($10^6 m^3$)	Total agricultural sector	83	80	80
	Studied crops	30	27	27
	Reduction with respect to studied crops (%)	0	10	10
	Reduction with respect to total (%)	0	4	4
Added value (10^6US$)	Total agricultural sector (PCBS, 2006)	267	277	270
	Studied crops	26	36	29
	Increase with respect to studied crops (%)	0	38	12
	Increase with respect to total (%)	0	4	1
Water value	Value ($/m^3$)	3.2	3.5	3.4
	Percentage increase (%)		9	6
*Scenario 1 represents the existing situation, Scenario 2 represents maximizing profit under water and land constraints and Scenario 3 represents maximizing profit under water, land and local consumption constraints.				

4.3.2 Cropping patterns

The results of modeling of scenario 2 suggest that in Jenin area the most feasible crop is eggplant, eggplant and citrus are feasible in Jericho, open irrigated squash and citrus in Nablus, in Tulkarem area open irrigated squash is the only crop feasible while tomatoes is suggested in Tubas area. For cucumber the model suggests that there is no need to cultivate it under open irrigation because the quantity cultivated in green houses is enough. Under scenario 3 extra area was allocated to tomatoes and rain-fed squash in Jenin, in Jericho the eggplant area was reduced with an extra area for tomatoes, squash and citrus, in Tulkarem and Tubas extra area was allocated for citrus, rain-fed squash was suggested in Tubas too. Table 4.4 presents example of cropping patterns in Jenin and Jericho zones.

Table 4.4 Existing and proposed area in (hectare) cultivated by each crop in the Jenin and Jericho zones under the 3 scenarios

Zone	Jenin			Jericho		
Crop*	Scenario 1	Scenario 2	Scenario 3	Scenario 1	Scenario 2	Scenario 3
Tomatoes, Op	145	0	625	371	0	365
Tomatoes, Gh	70	70	70	30	30	30
Cucumber, Op	511	0	0	190	0	0
Cucumber, Gh	169	169	169	46	46	46
Eggplant, Op	142	360	0	496	1600	724
Squash, Op	251	0	0	692	0	322
Squash, Rf	265	0	703	0	0	0
Citrus, Op	15	0	0	68	216	300
Total	1568	599	1567	1893	1892	1787
Scenario 1 presents the existing situation, scenario 2 presents the maximizing profit under land and water availability constraints and scenario 3 presents maximizing profit under land, water availability and local consumption constraints						
*Op stands for open irrigation, Gh for green houses and Rf for rain-fed						

4.3.3 Production

Table 4.5 presents the production of each crop under the three scenarios, We can see that in scenario 2 the production of tomatoes and citrus does not satisfy the West Bank demand and is less than the existing production while the production of eggplant is more than the existing production. In scenario 3 the demand is satisfied on the expense of profit.

Table 4.5 Total production of each crop under the three scenarios and the West Bank demand for each crop

Crop		Scenario 1	Scenario 2	Scenario 3	Demand for the crop in ton (1000 kg)
Production quantity in ton (1000 kg)	Tomatoes	83,560	76,152	81,000	81,000
	Cucumber	79,733	42,372	42,372	35,000
	Eggplant	39,358	100,899	36,182	17,000
	Squash	24,734	18,351	20,000	20,000
	Citrus	30,062	13,373	37,000	37,000
Scenario 1 presents the existing situation, scenario 2 presents the maximizing profit under land and water availability constraints and scenario 3 presents maximizing profit under land, water availability and local consumption constraints					

4.4 Discussion

4.4.1 Water use, productivity and profit

A reduction of 4% in the water use was achieved in both scenarios 2 and 3 with an increase of profit by 4% and 1% in scenario 2 and scenario 3 respectively. The water productivity was increased by 9% in scenario 2 and by 6% in scenario 3. The results of this study compare well with that of Al-Weshah (2000) where 14 vegetable crops cultivated in the Jordan valley were studied. He found a reduction of water use of 0.8% accompanied by a 5% increase in profit when the objective function was to maximize revenue, and a reduction of water use of 9% with a 2% increase in profit when the objective function was to minimize the water use. The water productivity was increased by 5% and 9% (Al-Weshah, 2000).

It can be noted that the marginal or average productivity of water varies according to cropping patterns used. Marginal water productivity can be used to investigate how to allocate water to the most beneficial use of water in terms of choosing the appropriate cropping patterns. Some may go further for trade-of among sectors' (agricultural, industrial, tourism) water uses based on their water use revenue, that is to allocate water for its most beneficial use (Al-Weshah, 2000).

4.4.2 Crop Patterns

In scenarios 2 and 3 the model has allocated a considerable area for rain-fed squash in Jenin and Tubas. This means that it is feasible and profitable to cultivate squash under the rain-fed irrigation system in both areas, thereby enhancing rain-fed cropping patterns as a means of reducing water use. Farmers should be motivated to shift from irrigated to rain-fed agriculture. According to PCBS (2006), many vegetable crops cultivated in the West Bank are feasible under rain-fed irrigation systems such as olives, grapes, wheat and many others. In this context, Falkenmark (2007) said that future food production will have to benefit maximally from rainfall rather than from irrigation. She added that climate data show that there is, also in semiarid regions, generally enough rainwater during the rainy season to meet consumptive water requirements. Yang *et al.* (2006) argued that the rain-fed agriculture has lower opportunity costs and environmental impact in terms of water use than irrigated agriculture. Soil moisture which is used for rain-fed crops is a free good in terms of supply; other plants are the only major competitive users of this water. This makes rain-fed agriculture attractive in terms of cost and environmental impact although the yield of rain-fed crops is less than that of irrigated crops. More research is needed to improve the yield of rain-fed crops. Greater efforts in terms of agricultural technology and investment should be devoted to the development of rain-fed agriculture. Given the increasing scarcity of global water resources, more effectively utilizing water stored in soil may also have to be a goal that world agriculture pursues in the years to come (Yang *et al.*, 2006).

The model suggested that each agricultural zone should be specialized in one or two crops from the five crops included in the model to ensure a maximum profit on the national scale. One may argue about the farmers' acceptance of such a distribution of crops within districts because of the lack of crop diversity within each district. Farmers have a variety of reasons for rejecting monoculture (planting one single crop over a wide area). On-farm needs for additional crops and the possibility of crop failure due to disease or insects attack are reasons why more than one type of crop is often planted on each farm (Loucks *et al.*, 1981). Some may even go further for inter-planting for pest control and insect confusion. For example, marigolds and sunflowers are a good choice for attracting helpful insects because of their wide open flowers. Herbs like parsley and thyme have strong fragrances that attract beneficial insects. Constraints ensuring at least some pre-specified crop diversification can be added to the model (Loucks *et al.*, 1981). Crops which are not included in the model still can contribute to ensure diversification.

In scenario 2 the model increased the area to be cultivated by eggplant in Jericho and Jenin and decreased the area of tomatoes. This means that eggplant should be planted in surplus and could be exported and in contrast tomatoes could be imported to satisfy the local needs. One may however argue in favor of self-sufficiency in terms of vegetable production. In scenario 3 a constraint was included to investigate the effect of self-sufficiency of the studied crops on profit. It was found that the profit decreases because some crops have to be cultivated to satisfy the demand while these crops are not profitable such as tomatoes and citrus. It appears that the optimal cropping pattern in Scenario 3 is closer to the existing cropping pattern than the profit-maximizing cropping pattern. Some may wonder why farmers are currently not cultivating profit-maximizing combination of crops suggested in scenario 2. This may be because farmers are driven by the self-sufficiency principle rather than exporting profitable crops. Moreover, exporting crops poses the risk of closures of exit terminals which are controlled by Israelis and could be closed for several days or even weeks which will cause significant loss to farmers as well as to the overall economy. A good example is what had happened in Gaza Strip in June 2007 when Israelis prevented the farmers from exporting strawberry and flowers which they have planted for export to the European Union. This caused some US$ 25 million loss to the farmers and the economy at that time (Cohen, 2007).

Decision makers should compromise between profit making via exporting and self-sufficiency. This brings the concept of virtual water trade into the picture; that is, importing water intensive and low valued crops and exporting high valued crops. In line with this Hoekstra and Chapagain (2007) and Qadir et al. (2007) stated that virtual water trade, through food trade, is an option towards dealing with water scarcity in countries with scarce water resources. Hoekstra and Hung (2005) stated that a water scarce country can aim at importing products that require a lot of water (water intensive products) and exporting products that require less water (water extensive products). This will relieve the pressure on the nations own water resources. A number of arid countries such as Jordan have consciously formulated policies to enable water saving by reducing export of water intensive products, notably crops. The remaining virtual water export is largely related to crops that yield relatively high income per cubic meter of water consumed (WWC, 2004). Al-Weshah (2000) argued that although virtual water trade is a means of water saving in water scarce countries, it poses the risk of creating job loss in the agricultural sector. Planners and policy makers should consider projects to shift activities in the same area. Al-Weshah (2000) added that many voices in all the countries sharing the Jordan River are calling for better water resources management in the agriculture sector. The calls of experts from Jordan and Israel suggested that importing some agricultural products may be more rational than producing them locally in terms of their water use. Food staples are commonly available at cheaper prices compared to the cost of producing them domestically for many water-poor nations.

4.4.3 Sensitivity analysis

Table 4.6 presents the sensitivity report in relation to constraints for scenarios 2 and 3. It presents the final value of each constraint and the shadow price. Regarding the area constraint in scenario 2, the shadow price of the area in Jericho was the highest US$ 7370/hectare which means that by increasing the area in Jericho by one unit (1hectare) the profit will increase by US$ 7370. For Tulkarem and Jenin, no additional profit is expected by increasing the area by the same amount as the shadow prices are zero (Table 4.6). The shadow price is important for decision makers in that, if the decision is about land reallocation for agriculture in order to increase profit, more land should be allocated for agriculture in Jericho followed by Tubas because these areas have the highest shadow prices.

The highest shadow price in relation to the water constraint in scenario 2 was US$ 1.6/m^3 in Jenin followed by Tulkarem area (US$ 1.46/m^3), this means that by increasing the water availability in Jenin and Tulkarem by one unit of water, the profit will increase by US$ 1.6 and US$ 1.46 respectively (Table 4.6). In Jericho, Nablus and Tubas the shadow prices are low which means that increasing the availability of water in these areas is not a wise decision if the objective is to increase profit. In this context Kumar and Khepar (1980) indicated that with linear programming models it is possible to choose among crops, when water becomes scarcer or the cost of irrigation water increases. The farmers may adjust the cropping patterns by decreasing the area under crops that demand more water.

In scenario 3 a negative shadow price was seen with respect to tomatoes, squash and citrus products. This means that by increasing the production of these crops by one unit (1ton=1000kg) a decrease in profit equal to the shadow price will be noticed. Therefore, it is more profitable to import these products rather than producing them domestically (Table 4.6). Shadow prices associated with water balance constraints are important to decision makers because they provide the marginal value of water related to the value in use which can help determining the areas where the value of water is high and the ranges of water demand within which this value is valid.

Table 4.6 Shadow price for water, land and production under Scenarios 2 and 3

Constraint	Scenarios 2		Scenarios 3	
	Final Value	Shadow Price	Final Value	Shadow Price
Area constraint	(hectare)	(US$/hectare)	(hectare)	(US$/hectare)
Jenin	599	0	1568	960
Jericho	1893	7370	1787	0
Nablus	469	2780	469	5110
Tulkarem	529	0	538	0
Tubas	846	3290	846	6190
Water constraint	(m^3)	$(US$/m^3)$	(m^3)	$(US$/m^3)$
Jenin	3,900,000	1.6	3,900,000	4.5
Jericho	12,000,000	0.03	12,000,000	1.5
Nablus	3,800,000	0.13	3,800,000	1.0
Tulkarem	4,100,000	1.46	4,100,000	3.9
Tubas	3,500,000	0.07	3,500,000	0.9
Local consumption constraint			Ton(1000kg)	(US$/ton)
Citrus			37,000	-485
Squash			20,000	-193
Eggplant			36,182	0
cucumber			42,372	0
Tomatoes			81,000	-170

4.5 Conclusions

Within the limitations of the research the following conclusions were drawn:

1. Changing the cropping patterns within the country can be used as a tool to reduce the agricultural water use. Moreover, it can simultaneously increase the added value per unit of water.
2. The linear programming model showed that a reduction of 4% of the total agricultural water use in the West Bank can be achieved by changing cropping patterns.
3. Applying the linear programming model under land and water availability constraints showed that the profit from studied crops can be increased by 38%. This would mean a 4% increase of the total agricultural sector. When the model was applied under the water and land availability and local consumption constraints to ensure self-sufficiency, the increase in the added value of the studied crops was 12%, equivalent to a 1% added value for the entire agricultural sector in the West Bank.
4. Expanding rain-fed agriculture is an effective method to address water scarcity; it can substantially reduce the agricultural national water consumption.

4.6 Recommendation

This model has been developed to provide a tool that can be used by decision makers in order to determine the optimal mix of crops in the West Bank that ensure the maximum profit on a national level. The model can be modified according to changes in allocation of water, land and any other related parameter. Also the model can be expanded to include other crops. This model has been formulated on the West Bank level taking into account five different agricultural districts in the West Bank. Models can also be formulated at district level if these districts have different agricultural zones.

Acknowledgment

Thanks are extended to the team of questionnaire fillers leaded by Nida' Shahbari and to all farmers who participated in the study. Thanks are also extended to engineer Qabas Bshara for his assistance and engineer Marwan Mokhtar for his help in processing the datat.

References

Allan, J.A., (1997) "Virtual water": A long term solution for water short Middle Eastern economies? Water Issues Group, School of Oriental and African Studies (SOAS), University of London, Paper presented at the 1997 British Association Festival of Science, Water and Development Session, 9 September 1997.

Al-Weshah, R.A., (2000) Optimal use of Irrigation Water in the Jordan Valley: A Case Study, Water Resour. Manage. 14, 327-338.

Cohen, A., (2007) Israel okays renewal of flower, strawberry exports from Gaza, Haaretz Newspaper, Israel. Available online http://www.haaretz.com/hasen/spages/926607.html. accessed Feb 2009.

Doorenbos, J., Kassam, A. H.,(1979) *Yield response to water.* FAO Irrigation and Drainage Paper 33, Food and Agriculture Orgnization, Rome, Italy.

Falkenmark, M.,(1986) Fresh water- time for a modified approach, Ambio 15(4), 192-200.

Falkenmark, M., (2007) Shift in Thinking to Address the 21st Century Hunger Gap, Moving Focus from Blue to Green Water Management, Water Resour. Manage. 21, 3-18.

Haouari, M., Azaiez, M. N., (2001) Optimal Cropping patterns Under Water Deficits, EUR. J. Oper. Res. 130, 133-146.

Hillier, F.S., Liebermam, G. F., (2005) Introduction of Operation Research, 8. McGraw-Hill, New York.

Hoekstra, A. Y., Hung, P.Q., (2005) Globalization of Water Resources: international water flows in relation to crop trade, Global Environ. Chang.15, 45-56.

Hoekstra, A. Y., Chapagain, A. K., (2007) Water footprint of nations: Water use by people as a function of their consumption pattern, Water Resour. Manage. 21, 35-48.

Karamouz, M., Szidarovszky, F., Zahraie, B., (2003) Water Resources System Analysis, Lewis publishers.

Loucks, D. P., Stedinger, J.R., Haith, D.A., (1981) Water Resource System Planning and Analysis, Prentice-Hall,Inc. Englewood Cliffs, New Jersey.

Mimi, Z. A., (2001) Water demands: modeling approaches, European Water Management. 4(2), 39-43.

Nazer, D.W., Siebel, M.A., Mimi,Z., Van der Zaag, P., Gijzen, H.J., (2008) Water footprint of the Palestinians in the west Bank, Palestine, J. Am. Water Resour. As. (JAWRA) 44(2), 449-458.

Oslo II Agreement, (1995) Israeli-Palestinian Interim Agreement on the West Bank and the Gaza Strip, Annex III, Article 40, Washington D.C., September 28 1995.

PCBS (Palestinian Central Bureau of Statistics), (1999) Population in the Palestinian Territory 1997-2025 , Palestinian Central Bureau of Statistics, Palestine.

PCBS (Palestinian Central Bureau of Statistics), (2006) Agricultural Statistics 2004/2005, Palestinian Central Bureau of Statistics, Palestine, available on line, http://www.pcbs.org

PCBS (Palestinian Central Bureau of Statistics), (2007) The Palestinian Expenditure and Consumption Survey (2006) Levels of Living in the Palestinian Territory, The Final Report (January 2006-January 2007), Palestinian Central Bureau of Statistics, Palestine.

PCBS (Palestinian Central Bureau of Statistics), (2008) Population, Housing and Establishment Census 2007, Census Final Results in the West Bank, Summary (Population and Housing), Ramallah- Palestine.

PWA (Palestinian Water Authority), (2004) Data collection by personal communication from the PWA data base, Resources and Planning Department, Palestinian Water Authority.

Playan, E., Mateos, L., (2006) Modernization and Optimization of Irrigation Systems to Increase Water Productivity, Agric. Water Manage. 80, 100-116.

Qadir, M., Sharma, B. R., Bruggman, A., Choukr-allh, R., Karajeh, F., (2007) None conventional water resources and opportunities for water augmentation to achieve food security in water scarce countries, Agric. Water Manage. 87, 2-22

Rogers, P., Bhatia, R., Huber, A., (1998) Water as a social and Economic Good: How to put the principle into practice, Global Water Partnership, Technical Advisory

Committee. Swedish International Development Cooperation Agency, Stockholm, Sweden, pp. 10-14.

Rogers, P., Silva, R., Bhatia, R., (2002) Water is an Economic Good: How to use prices to promote equity, efficiency and sustainability, Water Policy 4, 1-17.

Rosegrant, M. W, Cai, X., Cline, S. A., (2002) Global Water Outlook to 2025, Averting an Impeding Crisis, Food Policy Report, A 2020 Vision for Food, Agriculture, and the environment Initiative, International Food Policy Research Institute (IFPRI) and International Water Management Institute (IWMI), Colombo, Sirilanka.

Sunantara, J.D., Ramirez, J.A., (1997) Optimal Stochastic Multicrop Seasonal and Itraseasonal Irrigation Control, J. Water Res. PL_ASCE. 123(1), 39-48.

Vedula, S., Mujumdar, P.P., (1992) Optimal Reservoir Operation for Irrigation of multiple crops, Water Resour. Res. 28 (1), 1-9.

Vedula, S., Kumar, D.N., (1996) An Integrated model for Optimal Reservoir Operation for Irrigation of multiple crops, Water Resour. Res. 32(4), 1101-1108.

Walsh, G.R., (1985) An Introduction to linear Programming, 2, Department of mathematics, University of York. A Wiley- Interscience Publication, John Wiley and Sons.

World Water Council (WWC), (2004) E- conference Synthesis , Virtual Water Trade – conscious choices. WWC publication No.2.

Yang, H., Wang, L., Abbaspour, K., Zehnder, A., (2006) Virtual water and the need for greater attention to rain-fed agriculture, Water 21, Magazine of the International Water Association, pp 14-15.

Chapter 5

Saving Water and Reducing Pollution in the Unhairing-Liming Process in the Leather Tanning Industry

Based on:

Nazer, D. W., Al-Sa'ed, R. and Siebel, M. A. (2006) Reducing the Environmental Impact of the Unhairing-liming Process in the Leather Tanning Industry, Journal of cleaner production volume 14 (1), pp 65-74.

Saving Water and Reducing Pollution in the Unhairing-Liming Process in the Leather Tanning Industry

Abstract

The tanning industry is classified as a high environmental impact industry due to the high concentrations of organics, salts and heavy metals (chromium) used, the high concentration thereof in the wastewater and the large quantities of solid waste generated in this industry. Specifically water use in this industry is significant, typically exceeding 50 m^3 per ton (1000 kg) of hides processed in the blue wet process. This study aims at reducing the water consumption and the environmental and financial impacts of the leather tanning industry by introducing a modified unhairing-liming treatment that reuses the water and -partly- the chemicals. Experiments were carried out at lab scale. Life cycle impact assessment was used to evaluate the environmental and economic benefits of the proposed treatment. The quality of the produced leather was assessed by experts from the tanning sector (tanners). It was concluded that the unhairing-liming process water could be reused four times without affecting the quality of the final leather product. The proposed unhairing-liming treatment permits savings in water consumption and wastewater production of up to 58%. The chemicals consumption was reduced; sodium sulfide by 27% and lime by 40%. Moreover, there is a reduction of the overall environmental impact with up to 16%.

Key words: Financial impact, reducing environmental impact, reduction of chemicals, tanning industry, unhairing-liming water reduction, water reuse.

5.1 Introduction

Global water resources are limited and so are those in the West Bank, Palestine. The already existing water shortage in the area is expected to grow more serious in the near future as water consumption is increasing (MOPIC, 1998). Moreover, the water resources are threatened by water pollution due to the inadequate wastewater treatment and disposal. This decreases water quality and, therefore, availability of good quality water. As a result, there is an urgent need for Palestinians to manage their scarce water resources adequately so as to change the present situation of increasing water scarcity into a situation of sustainable use of water.

In water sustainability studies, all water uses should be studied. In the West Bank, the industrial water use is small relative to other water uses, but exact figures are not available. Industrial consumption is usually included in the domestic water consumption. Together industrial and domestic comprise some 30% of the total water use.

The industrial sector is considered a main environmental polluter. The tanning industry is one example of a high water consuming and heavy polluting industry (UNEP, 1994; Chernicharo and Vliet, 1996). The water consumption in the tanning industry varies significantly from tannery to tannery and from hide to hide, and is usually in the range of 25-80 m^3 of water per ton of raw hides (UNEP, 1994). A corresponding amount of wastewater is produced. Because the various process streams are usually mixed, the composition of the tannery effluent is very complex and characterized by high pollutant concentrations. For example, in the hair pulp/chrome tanning operation, chemical oxygen demand (COD) concentration in the wastewater may reach a level of 27000 mg/l. The concentration of sulfide in the wastewater may reach 200mg/l. Moreover, the amount of solid waste produced in this industry is significant, for each ton (1000kg) of raw salted hides processed between 680 to 850 kg of solid waste is produced. With other words to produce 1 ton of leather between 2 and 6 tons of solid waste is generated (UNEP, 1994). In Palestine there are 18 tanneries. The water consumption of these tanneries is estimated at 0.07 Million m^3/year.

The objective of this study is to reduce the water consumption, environmental impact and the production cost of the unhairing-liming process in the leather tanning industry. The conventional process of unhairing-liming uses, for each new batch of hides or skins, fresh water and chemicals according to the standard recipe and discharges the used process water. This study proposes a method for unhairing-liming of hides that uses fresh water and chemicals for the first batch of hides or skins but that subsequently reuses this water and these chemicals for subsequent batches of hides or skins. Consequently reduced quantities of water and chemicals are used.

5.2 Background
5.2.1 Cleaner production and its opportunities in leather tanning
The term Cleaner Production was developed by an expert working group in 1989 as advice for UNEP's Industry and Environment Program. It is commonly defined as *"the continuous application of an integrated preventive environmental strategy applied to processes, products and services to increase overall efficiency and reduce risks to humans and the environment"* (UNEP, 1999). Cleaner production refers usually to improving industrial production processes so as to reduce the flow of waste products and, therefore, to reduce the impact thereof onto the environment (Gijzen, 2001; Siebel and Gijzen, 2002). Often waste streams are mixed which complicates material recovery. Waste is discharged without realizing its value as a raw material (Gijzen, 1998; Siebel and Gijzen, 2002). Cleaner production provides ways to reduce the industrial impact on the environment through pollution prevention rather than end-of-pipe treatment (UNEP, 1995; Siebel, 1999).

Most of the leather manufacturing steps take place in aqueous medium. The process water is usually disposed off as wastewater. Recycling the process water in the different steps of the leather manufacturing is a potential opportunity for reducing environmental impact. Sendic (1995) stated that the introduction of discontinuous washing and recycling of process water in the beam house reduces water consumption from 35 m^3 to 7 m^3/ton of hides without additional technological changes. A 25-40% reduction of COD was recorded. Similarly Khan *et al.* (1995) said that some of the unhairing-liming techniques permit a direct reuse of the process water after decantation to separate the sludge containing lime, fat and protein and the recharging with chemicals according to recipes. The procedure saved water, sodium sulfide and lime and reduced the COD load in the effluent.

5.2.2 Leather tanning
Animal skin structure
The skin is of fundamental importance for the animal. It acts as a tough, flexible and preventive membrane to help the animal control form, shape and size and make direct contact with the external environment. Animal skin consists of three main layers epidermis, dermis and hypodermis. *Epidermis* is the outer most layer. *Dermis* is the middle layer and has two fairly distinct regions, the papillary layer and the fiber network layer. The fiber-network layer is the main region of dermis and usually forms the major part of leather. *Hypodermis* is the lower layer. Only the middle layer (dermis) is used in leather making. In the operation called fleshing much of the lower layer (hypodermis) is removed, epidermal structures are loosened chemically and removed to a large extent in the process of unhairing (Reed, 1972).

Leather making process
Tanning is the process of converting hides and skins into leather. The process comprises the use of water and chemicals and consists of three main stages:
- Preparation of hides in the beam house.
- Tanning.
- Post tanning and finishing.

The quantity of water used through the leather making process varies significantly from tannery to tannery, from country to country and according to the type of skin or hide processed (bovine hides, sheep skins, etc.) (UNEP and UNIDO, 1991; UNEP/IE, 1996). Figure 5.1 shows average values for the volumes of water used per ton of hides in the various process steps (UNEP and UNIDO, 1991). These volumes were used for calculation purposes in the subsequent sections of the paper.

Beam house stage
Soaking: the purpose of soaking is to re-hydrate the hides, to remove the salt and blood and to facilitate the removal of nonstructural (non-collagenous) proteins and dirt in order to prepare the hides for further treatment (UNEP/IE, 1996). *Unhairing-liming:* The purpose of unhairing-liming is to remove the hair and epidermis. The hides are treated with 3% sodium sulfide containing 25% sulfide and 3% hydrated lime (calcium hydroxide) in a 200% float (a solution consisting of 2 tons of water per ton of skins or hides processed) (UNEP and UNIDO, 1991; UNEP/IE, 1996). The sulfide pulps the hair and the epidermis. The lime is used as a buffer to keep the pH at about 13 which causes the hides to swell, the collagen fiber network to open, and which helps the removal of the nonstructural proteins (UNEP/IE, 1996).

Tanning stage
In the tanning stage the hides are treated with an agent that displaces the intercellular water and combines with the collagen fibers. This increases their resistance to heat, hydrolysis and microbial degradation. Tanning stabilizes the collagen; the tannins join with the skin protein to form leather. Chromium salts are the most widely used tannins, although vegetable and synthetic tannins are also used (Khan *et al*, 1995).

Post-tanning and finishing

After tanning some mechanical operations may be done to level the surface of the material (UNEP, 1994). Among these operations are *sammying* which is used to remove excess moisture, *splitting* is the process of dividing the material into two layers (sometimes is carried out in the lime condition), *shaving* whereby the material is leveled yielding a waste of small fragments and *post-tanning wet work,* this involves further processing of the stabilized collagen network and may comprise a further tannage. After the post-tanning processes, the hides are dried and pressed, to create the correct shape and surface texture. They may also be trimmed, split and shaved. A decorative and protective surface coating may also be applied (UNEP, 1994).

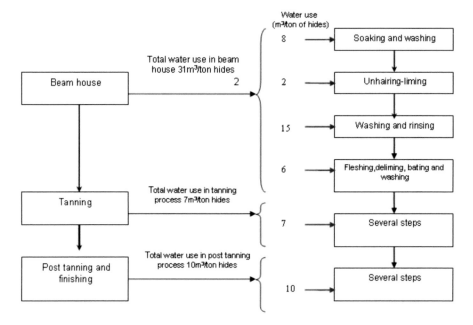

Figure 5.1 Scheme of the leather making process indicating the water use (per ton of raw hides) in each stage for processing cattle hides (adapted from UNEP and UNIDO, 1991).

5.2.3 Life cycle impact assessment (LCIA)

Life cycle impact assessment (LCIA), explained in chapter 3, is the process of evaluating the impacts of products, processes and services over the entire period of their life cycle. It involves the evaluation of the impact of the three dimensions of sustainable development; environmental, financial and social.

The environmental life cycle impact assessment (E-LCIA) is focusing on quantifying the environmental impacts of a product, process or service. Financial life cycle impact assessment (F-LCIA) involves the quantification of the financial impacts of a product, process or service during all phases of its life cycle (production, operation and disposal) taking into account one-time cost, recurrent costs and revenues as well as the interest and inflation. Social life cycle impact assessment (S-LCIA) is the process of quantifying the social impacts of a product, process or service over its life cycle.

5.2.4 Statistical methods
Chi-square test can be used for testing independence of two characteristics by calculating the expected frequencies assuming independence and the observed frequencies. The calculated observed value of the Chi-square will be compared with the critical value of Chi-square under the required level of significance to determine whether the two characteristics are related (Chase and Bown, 1986).

T-test is one of the statistical tests that permits making generalization or drawing conclusions about the attributes of a population from evidence contained in a sample (usually small samples <30) (Chase and Bown, 1986; Bohrnstedt and Knoke, 1994). The t-test can be used to compare between two types of treatments applied to two different independent or dependent (paired) samples by investigating the difference between the means of the two samples and to draw conclusions about the treatments (Chase and Bown, 1986; Bohrnstedt and Knoke, 1994).

5.3 Materials and methods
5.3.1 Experimental approach
The experimental work was realized at Birzeit University. The New Tanning Company in Hebron provided the raw hides and all chemicals needed for processing these hides. Three types of treatments were realized in this study:

Treatment 1	In this treatment the once-through approach was used, that is, fresh water with chemicals according to recipe (2 m^3 of water, 30 kg sodium sulfide and 30kg lime per ton of hides) was used to process the raw hides. The process water was disposed off after treating the hides (Figure 5.2 A). This treatment is equivalent to the conventional one used in the tanneries in Palestine and it was the reference against which treatments 2 and 3 were compared.
Treatment 2	The use-treat-reuse approach was used, that is, the water was used and once reused. In the first use, fresh water and chemicals were used according to recipe of treatment 1. Then process water of this first use was decanted, the volume and chemicals concentrations adjusted and reused to process another batch of hides (called recycle1). In this approach reduced volumes of fresh water and reduced quantities of chemicals were used (Figure 5.2 B).
Treatment 3	In this treatment the water was used five times. Fresh water and chemicals were used in first use according to the recipe of Treatment 1. That is recycle 1, 2, 3 and 4 were treated in the same way as recycle 1 in treatment 2 (Figure 5.2 B).

Experiments for treatments 1 and 2 were carried out 8 times while experiments for treatment 3 were carried out 4 times.

Decanted process water: Before being used in Treatments 2 and 3, the process water was left standing for 1 hour in order for the particulate matter to settle. Then the clarified process water was separated from the sludge. The volume of the decanted process water was indicative of the volume of replenishment water. Subsequently, the reduced quantity of chemicals was determined as follows (Figure 2): According to UNEP (1994) the effluent of the unhairing-liming will still contain 50% of the sulfide and 66% of the lime. The volume of the process water that can be reused was found to be 1.4 m^3 out of 2 m^3 fresh water originally used to process 1 ton of hides. Therefore, this amount of water is estimated to still contain about 10kg of sulfide and almost 13kg of lime. In order to make up for the spent sulfide another 20 kg of sodium sulfide was added. Regarding lime, in recycle1 of Treatment 2, 20 kg was added. Afterwards it turned out that 15 kg of lime is adequate to be used in treatment 3 (Figure 5.2B).

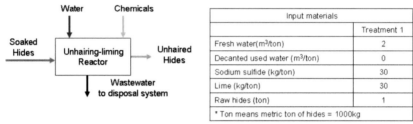

Input materials	
	Treatment 1
Fresh water(m^3/ton)	2
Decanted used water (m^3/ton)	0
Sodium sulfide (kg/ton)	30
Lime (kg/ton)	30
Raw hides (ton)	1
* Ton means metric ton of hides = 1000kg	

A. Conventional method of unhairing-liming Treatment 1 (once-through approach)

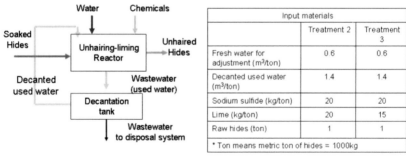

Input materials		
	Treatment 2	Treatment 3
Fresh water for adjustment (m^3/ton)	0.6	0.6
Decanted used water (m^3/ton)	1.4	1.4
Sodium sulfide (kg/ton)	20	20
Lime (kg/ton)	20	15
Raw hides (ton)	1	1
* Ton means metric ton of hides = 1000kg		

B. Modified method of unhairing-liming Treatment 2 and 3 (use-treat-reuse approach)

Figure 5.2 Schematic diagrams of the unhairing-liming method: part A presents Treatment 1 and part B presents Treatments 2 and 3.

5.3.2 Experimental apparatus

An experimental apparatus was designed, assembled and operated to simulate a real tannery process. The apparatus consisted of three reactors of the same size. The first acted as a soaking reactor, while the second and third acted as unhairing-liming reactors.

The unhairing-liming reactors were running off the same engine and timer so as to ensure identical operating conditions in terms of the rotational speed (Figure 5.3).

Figure 5.3 Overview of the three reactors.

5.3.3 Materials
Water: Fresh water was used for the soaking step and the unhairing-liming step in Treatment 1 and the first use of Treatments 2 and 3. Fresh water was also used for the adjustment of the water quantity in the remaining recycles of Treatments 2 and 3.
Sodium sulfide and hydrated lime: The hides were treated with sodium sulfide containing 25% sulfide and hydrated lime, quantities are given in the experimental approach paragraph.
Hides: Wet salted bovine cattle hides were used.

5.3.4 Experimental methods
Day 1 Pieces of hides were prepared of equal size, 30*30 cm^2. The exact sizes and weights of the pieces of hide were recorded and the pieces were marked so as to differentiate between pieces processed by the three types of treatment. The pieces of hide were washed to remove dirt and manure and then soaked for 24 hours in the soaking reactor with 200% float (weight of fresh water as percentage of weight of hides).
Day 2 The soaked pieces were transferred to the unhairing-liming reactors. Some pieces were processed according to Treatment 1; others were processed according to Treatments 2 or 3. Treatments 1 and 2 were repeated eight times while Treatment 3 was repeated 4 times.
Day 3 The unhaired pieces were unloaded and new pieces were processed in the same way. Each time the reactors were unloaded, the effluent of each reactor was analyzed the same day for Sulfide S^{-2}, Sulfate SO$_4^{-2}$ and COD (Three samples each). The unhaired pieces of hide were further processed in the tannery to produce the final product of the pieces of hide (now called pieces of leather).

5.3.5 Life cycle impact assessment

System boundaries were chosen just around the unhairing-liming process since processes preceding and following the unhairing-liming step are not affected by the proposed process modification. LCIA involves the three phases of a product, process or service *i.e.* production, operation and disposal. For practical reasons only the operational phase will be included in the environmental and financial calculations. Social impacts associated with the production process were virtually absent in this study, that is in both the original method and in the new methods, the job of the operators is almost the same. Therefore, social impacts were not included in the impact assessment.

Environmental life cycle impact assessment: The three treatments in this study were evaluated using the Eco-indicator 99 database (Goedkoop *et al.*, 2000) in which the environmental impact was given in eco-point/unit weight of material.

Financial lifecycle impact assessment: To evaluate the financial costs of the operational phase of reusing the unhairing-liming process water the following steps were carried out:
5. Determination of the annual operational costs (US$/year) of extra labor, electricity and maintenance.
6. Determination of the annual savings in operational costs from reduced costs of water and chemicals.
7. Determination of the present worth of the annually returning operational costs.
The present worth (PW) is one such method which relates the cost of any activity at certain time to the cost at another time given certain values for interest rate and inflation (Philippatos and Sihler, 1991; Blank and Tarquin, 2005). The PW can be calculated according to equation 3.1 given in chapter 3. In order to calculate the present worth for annual operational costs and benefits only, the investment is not included in the equation.

The above cost components may be limited to get the real picture; the cost of cleaning up, and polluter pays principle would add heavily on the benefit side of cleaner production interventions. However, this cost was not included in the calculations because the principle of polluter pays is not applied in Palestine.

5.3.6 Quality assessment

Pieces of leather processed in the three treatments and visibly identical were mixed and evaluated by 6 tanners. The tanners were asked to examine the pieces of leather and classify them according to quality.

5.3.7 Statistical methods

The average and standard deviation for COD, Sulfide and Sulfate concentrations in the effluent of processing each batch of hides were calculated for the different types of treatment. The chi-square test was used to determine whether there was quality difference between the pieces of hides processed in the three types of treatment. In order to determine the number of times the process water could be reused, first the lab work began with Treatment 1 and 2 that is applying one time reuse of process water, during this experiment the significance of the reduction of water consumption, COD, sulfide and sulfate emissions between the first use and recycle 1 of Treatment 2 was investigated using the t-test for paired samples. Finding out that there is a significant reduction in the above mentioned parameters, Treatment 3 was applied with 4 recycles (i.e. water was used 5 times). The significance of the reduction of water consumption, COD, sulfide and sulfate emissions between each recycle and the following one was again investigated using the t-test for paired samples in order to determine how may times the process water could be reused.

5.4 Results and discussion
5.4.1 Water consumption
Experiments were carried out with the aim to reduce water and chemicals consumption in the unhairig-liming step of the tanning process. Process water was
1.   used one time and discharged (Treatment 1) or
2.   used and one time reused (Treatment 2) or
3.   used and 4 times reused (Treatment 3).

The results of the study indicate that the average water consumption per ton of hides was reduced from 2 m^3 per ton of hides in Treatment 1which was the basis of the comparison, to 1.3 m^3 per ton of hides (36% reduction) in Treatment 2, to 0.8 m^3 per ton (58%) in Treatment 3 (Table 5.1). The overall savings in water consumption increased by increasing the number of times process water was reused (Table 1). These results compare well with Sendic (1995) who indicated that the water consumption in the beam house was reduced by 50% through recycling. The volume of the process water that could be reused after decantation was less than the input fresh water because the hides absorb part of the water, another part could not be used because it is mixed with the sludge after decantation. The volume of not reusable water was measured in each cycle. It was found that, on average 27% of the effluent could not be reused. Fresh water needed to be added with each next batch of hides.

Table 5.1 Average and standard deviation of water consumption and water savings per ton of hides for each use.

Treatment Type	Treatment 3			
	Treatment 2			
	Treatment 1			
	First use	Recycle 1	Recycle 2	Recycle 3	Recycle 4
Average water consumption (m^3 per ton of hides)	2	1.3±0.06	1.0±0.03	0.9±0.02	0.8±0.03
Water savings per ton of hide (m^3)	0	0.7±0.06	1.0±0.03	1.1±0.02	1.2±0.03
Number of experiments (N)	8	8	4	4	4

5.4.2 Environmental Impact of COD, sulfide and sulfate
Table 5.2 presents the environmental impacts, expressed in milli-Ecopoint per unit of product, from the release of sulfide, sulfate and COD for the three treatments. The environmental impact of the sulfide and sulfate were found zero, the substances do not

contribute to any global impact category (greenhouse effect, global warming, acidification, eutrophication, carcinogenesis. etc) although they may have quite a strong local impact (PRé Consultants, 2002). From Table 5.2 it can be concluded that the total environmental impact was reduced. The unhairing-liming process emits COD, sulfide and sulfate to water. COD is generated from the processing of the raw hides and from the chemicals used to process the hides while sulfide and sulfate are produced from the use of sodium sulfide in the process. Treatment 2 and 3 proposed in this research used a reduced quantity of chemicals thus reducing the emissions of COD, sulfide and sulfate. COD emission was reduced by 16% and 33% in treatments 2 and 3 respectively compared to treatment 1. Sulfide emission was reduced by 31% and 56%. In this context Sendic (1995) recorded a reduction of 20-40% in COD emission and 74-77% in sulfide emission when reusing of the process water was practiced. Aloy et al., (2000) and IULTCS (2004) also stated that the direct reusing of liming float could give a decrease of 30 to 40 % of COD. The reduction in COD and sulfide emissions depends on the quantity of sodium sulfide added for the process water as make up and the number of times the process water was reused.

5.4.3 Financial aspects

Savings of US$ 36 per ton of hides were achieved in the operation of the unhairing-liming process when implementing Treatment 2 and US$ 74 per ton of hides when implementing Treatment 3 (Table 5.3). Savings were achieved from a reduced consumption of water and chemicals as well as from reduced wastewater treatment costs because of a reduction in wastewater production. Moreover, less wastewater allows for smaller wastewater treatment plants (Sendic, 1995). UNEP (1994) showed that trials have indicated that between 20% and 50% saving of sulfide and 40% to 60% saving of lime could be achieved. Savings in the cost of sodium sulfide achieved in this study were 17% and 27% in Treatments 2 and 3 respectively. For lime the percentage was 20% and 40% respectively.

5.4.4 Relationship between environmental and financial impacts

Figure 5.4 shows the relationship between environmental and financial benefits. It can be seen that for one milli-Ecopoint decrease in the environmental impact there is a profit gain of US$ 1.8, this means that the more environmentally friendly the treatment is, the less costly it is. In contrast, Barrios et al. (2008) found a relationship between the financial consequences of further reducing the environmental impacts of water treatment processes of some 15 Euro-points per eco-point when they studied the environmental and financial impacts of different water treatment processes. This means that introducing environmentally friendly methods of treatment are not always more costly; it depends on the specific case and circumstances. Siebel et al. (2007) stated that, introducing environmental impacts in financial terms offers a way to investigate whether the more environmentally friendly approach is more costly.

Table 5.2 COD, sulfide and sulfate emissions in kg/ton of hides and total environmental impact in milli-Ecopoints/ ton of hides in the three treatments of the unhairing-liming process

Treatment		Treatment 1			Treatment 2			Treatment 3								
		First use			Recycle 1			Recycle 2			Recycle 3			Recycle 4		
Pollutant released	Environmental Impact/unit pollutant released	Q_e [a]	σ [b]	EI [c]	Q_e	σ	EI	Q_e	σ	EI	Q_e	σ	EI	Q_e	σ	EI
	mPt /unit															
COD	2.88	43	6.5	124	36	2.4	104	32	2.3	92	30	2.9	86	29	3.9	84
Sulfide	0.0	4.5	0.54	0.0	3.1	0.5	0.0	2.4	0.04	0.0	2.0	0.23	0.0	2.0	0.18	0.0
Sulfate	0.0	0.19	0.06	0.0	0.14	0.04	0.0	0.10	0.02	0.0	0.05	0.01	0.0	0.05	0.01	0.0
Energy	1.88	70		132	70.1		132	70.2		132	70.3		132	70.4		133
Total				256			236			224			218			216
Number (N) [d]		24			24			12			12			12		

[a] Q_e = mass released (kg/ton of hide) or energy (kW/ton of hides).

[b] σ =Standard deviation

[c] EI = environmental impact (mPt/ton of hide) (mPt =milli-Ecopoints).

[d] N = Total number of samples analyzed for COD, Sulfide and Sulfate. N=number of experiments multiplied by 3 samples.

Table 5.3 Annual operational benefits, costs and present worth of the three treatments.

Treatment	Treatment 1	Treatment 2	Treatment 3		
	First use	Recycle 1	Recycle 2	Recycle 3	recycle 4
Total annual benefits from water, wastewater treatment and chemicals savings (US$/year)	0.0	3485	4675	5250	5635
Total annual operational costs from extra labor, electricity and maintenance (US$/year)	0.0	1382	1382	1382	1382
Net annual benefits= Total benefits – total costs (US$/year)	0.0	2103	3293	3868	4253
Present v worth PW (US$/year)	0.0	9105	14259	16748	18415
Present worth PW per ton of hides (US$/ton of hide)	0.0	36	57	67	74

*Calculations were made for a tannery processing 250 ton of hides per year using the following information:

- Present worth was calculated according to the equation below:

$$PW = A \left[\frac{(1+k)^n - 1}{k(1+k)^n} \right] - I_0 \quad \text{(Blank and Tarquin, 2005)}$$

Where,

PW: Present worth, is the monetary value at present or at time zero; A: Net annual benefits; k: Dicount rate;

n: Number of years; I_0: Investment in year zero.

- Discount rate was estimated at 5% according to local conditions. (CIA,2009)
- Number of years was assumed five years (life of the equipment).

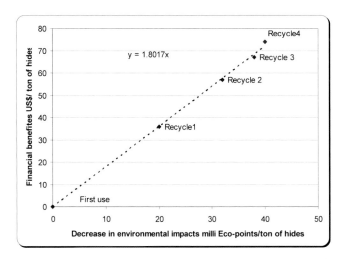

Figure 5.4 Relationship between the reduction of environmental impacts and the financial benefits achieved in the three treatments

5.4.5 Quality assessment

The pieces of leather processed in the three types of treatments were examined by the tanners who were asked to classify the pieces of leather according to quality.

Although there are a wide variety of tests possible for assessing quality of the produced leather, these tests are usually carried out in specialized laboratories disposing of specialized testing equipment. Such facilities are not available for the tanning industry in Palestine. In this study, the tanners used their expertise to evaluate the products. As such the procedure was subjective.

The main objective of the examination was to assess whether there was a quality difference between the pieces of hides treated in fresh water and those treated in used water. Using the results of the quality assessment by tanners, the Chi-Square test was used to check if the quality of the produced leather was related to the type of treatment. According to the tanners' classification and the Chi-square calculations, it was found that the quality of the final product of leather was not affected by the type of treatment: the unhairing-liming process water could be reused four times without affecting the quality of the produced leather. However, in case of quality deterioration during the process, this deterioration may be compensated for in the subsequent steps of leather processing as several authors stated that the quality of the leather produced might be affected negatively through direct reusing of unhairing-liming process water, but could be improved during the subsequent phases of leather processing (Aloy *et al.,* 2000; IULTCS, 2004). In this study four times reuse of the process water were found sufficient to produce good quality leather. This is far below the 10 times indicated by Cranston *et al.* (1989) who found that when the lime liquor in the SiroLIME[TM] hair saving process was reused more than 10 times; there was incomplete removal of epidermis.

5.4.6 Statistical explanation

From Tables 5.1 and 5.2, it can be seen that the extent of reduction of water consumption and environmental impact between recycles 3 and 4 is much smaller than that between the previous recycles. In fact, the impact reduction is approaching zero as the recycle number increases. The t-test was used to investigate the significance of the difference between each recycle and the following one. Regarding water consumption, the results of the test showed the t-value (34) between the first use of water and recycles 1 is larger than the critical one (1.895) at 5% level of significance which means that there is a significant reduction in the water consumption. The same observation can be made between recycles 1 and 2, 2 and 3, and 3 and 4; hence there is a significant reduction in the water consumption. Results of the paired t-test show no significant difference between recycle 3 and 4 with respect to the COD, sulfide and sulfate emissions (Table 5.4). According to these results one can conclude that although the quality of the leather is not affected due to the four times reuse of process water, recycle 4 of Treatment 3 is not justified with respect to the reduction of COD, sulfide and sulfate emissions.

Table 5.4 Paired t-values for water consumption, COD, sulfide and sulfate emissions (Chase and Bown, 1986).

Cycle	T-value				
	Water consumption	COD	Sulfide	Sulfate	Critical t-value *
First use, recycle1	34	2.4	19.7	6.1	1.895
Recycle 1, 2	12.7	4.1	3.9	2.4	2.353
Recycle 2, 3	8.3	3.5	3.7	3.4	2.353
Recycle 3, 4	4.3	0.5	0.3	1.4	2.353
*Level of significance 5%					

5.5 Conclusions

Within the limitations of this study, it can be concluded that: Although the process water of the unhairing-liming process could be reused four times without noticeable quality reduction of the final product of leather, reusing the process water for more than three times is not of significant benefit with respect to the reduction in COD, sulfide and sulfate emissions.

Acknowledgment

Special thanks are extended to Mr. Ramadan Za'tari, director general of the New Tanning Company in Hebron, Mr. Hani Al-Za'tari, and Mr. Talal Al-Za'tari for their dedicated cooperation and help in providing the necessary materials, and in processing the hides. Thanks are also due to Safwan Nazer and Jawad Nairoukh, for their help in preparing the apparatus.

References

Aloy, M., Fennen, J., Frendrup, W., Gregori, J., Ludvik, J., Money, C., Munz, K. H., Pantelaras, P., Rajamani, S., Toumi, A. and Van Vliet, M. (2000) IUE recommendations on cleaner technologies for leather production. Available at http://www.iultcs.org.

Barrios, R., Siebel, M., Helm, A., Bosklopper, K. and Gijzen, H. (2008) Environmental and Financial life Cycle Impact Assessment of Two Plants of Amsterdam Water Supply, Accepted for publication in Journal of cleaner production, volume 16(4), pp 471-476.

Blank, L. and Tarquin, A. (2005) Engineering Economy, 6th ed, McGrawHill.

Bohrnstedt, G. and Knoke,D. (1994) Statistics for Social Data Analysis, 3rd ed., F.E Peacock publishers, Itasca, Illinois.

Chase, W., and Brown, F. (1986) General statistics. John Wiley and Sons, Inc. New York.

Chernicharo, C. A. L. and Vliet, M. V. (1996) Strategies for pollution control in tanneries located in Minas Gerais state, Brazil case study, Wat. Sci. Tech. 34(11): 201-207.

CIA,   (2009)   CIA-The   world   fact   book   West   Bank. http://www.cia.gov//library/publications/the_world_factbook/geor/.html., accessed March 2009.

Cranston, R. W., Money, C. A. and Abbott, D. J. (1989) The use of the SiroLIME^Tm process in production of wet-blue leather (part-1 and part-2), Paper presented at IULTCS Congress, Philadelphia 1989. also available on line http://www.csiro.au.

EEA (European Environmental Agency)(1997), Life Cycle Assessment, A guide to Approaches, Experiences and Information Sources, available on line, retrieved October   4,   2007   http://reports.eea.europa.eu/GH-07-97-595-EN-C/en/Issue%20report%20No%206.pdf

Gijzen H.J. (1998) Sustainable wastewater management via re-use – Turning Waste into Wealth. In: Garcia, M., Gijzen, H.J., Galvis, G. (eds.), Proc. AGUA_ water and Sustainability, June 1-3, 1998, Cali, Colombia,pp 211-225

Gijzen, H. J. (2001) Aerobes, anaerobes and phototrophs – a winning team for wastewater management. Published in Wat Sci Tech. Vol 44, no 8, pp123-132.

Goedkoop, M., Effting, S. and Collignon, M. (2000), The Eco-indicator 99, A Damage Oriented Method for Life Cycle Impact Assessment, Manual for Designers, second edition, Pre consultants, Amersfroot, available on www.pre.nl. Accessed June 2008.

Hauschild, M., Jeswiet, J. and Alting, L. (2005) From Life Cycle Assessment to Sustainable Production: Status and Perspectives, Annals CIRP, 54(2), pp 535-555.

IULTCS (International Union of leather Technologies and Chemists Societies) (2004), IUE recommendations on Cleaner Technologies for leather production, updated document,   IUE   Commission.   Available   on   line   at http://wwww.iultcs.org/pdf/IUE.1-logo.pdf . Retrieved November 2007.

Khan, A. U., Iqbal, M., Ghous, M. R., Ul-Haq, I., and Bilgrami, R. (1995) International Cleaner Technologies in Tannery Clusters of Punjab, Pakistan Tanners Association (PTA) and Leather Industries Development Organization (LIDO).

MOPIC (Ministry of Planning and International Cooperation) (1998) Emergency Natural resources Protection Plan for Palestine "West Bank Governorates", Ministry of Planning and International Cooperation, Palestine.

Nazer, D. W., Al-Sa'ed, R. and Siebel, M. A. (2006) Reducing the Environmental Impact of the Unhairing-liming Process in the Leather Tanning Industry, Journal of cleaner production volume 14 (1), pp 65-74.

Philippatos, G. C., and Sihler, W.W. (1991) Financial Management text and cases, 2nd ed., Allyn and Bacon, Massachusetts.

PRé Consultants (2002) Personal communications through Maarten Siebel UNESCO-IHE, The Netherlands.

Reed, R. (1972) Ancient Skins Parchments and Leather, Department of Food and Leather Science, University of Leeds, England, pp13-85.

Sendic, V. M. (1995) Strategies in agro-industrial wastewater treatment. *Wat. Sci. Tech.* **32**:113-120.

Siebel, M. A. (1999) Cleaner Production-from environmental pollution to environmental responsibility, Paper presented at the 4th Princess Chulabhorn International Science Congress, Bangkok, Thailand.

Siebel, M.A. and Gijzen, H.J. (2002) *Application of cleaner production concepts in urban water management* in: Workshops Alcue–Sustainable development and urbanization: from knowledge to action, Science & Technology, Vol.3.

Siebel, M.A., Chang, CT., Gijzen, H.J. and Rotter, V.S. (2007) Life Cycle Impact Assessment- A Case Study of solid waste Collection, Proceedings of Eleventh international Waste Management and Landfill Symposium, S. Margherita di Pula, Cagliari, Italy; 1-5 October by CISA, Environmental Sanitary Engineering Centre, Italy.

Tapia, M., Siebel, M., Helem, A., Braas, E. and Gijzen, H. (2008) Environmental, Financial and Quality Assessment of Drinking Water Processes at Amsterdam Water Supply Company (Waternet), accepted for publication in Journal of cleaner production, volume 16(4), pp 401-409.

Udo de Haes,H., Heijungs, R., Suh, S., Huppes, G. (2004), Three Strategies to Overcome the limitations of Life-cycle Assessment, journal of Industrial Ecology, Volume 8, # 3, pp19-32.

UNEP (United Nations Environment Programme) and UNIDO (United Nations Industrial Development Organization) (1991) Audit and reduction Manual for Industrial Emissions and Wastes. Technical Report series # 7, United Nations publications, pp 56-71.

UNEP (United Nations Environment Programme) (1994) Tanneries and the Environment. A Technical Guide, Technical Report series # 4, United Nations Environment Programme, Industry and Environment/ Programme Activity Centre, United Nations publications. Paris, France, pp15-39.

UNEP (United Nations Environment Programme) (1995) Cleaner Production Worldwide. Volume II, United Nations Environment Programme, Industry and Environment, Cleaner Production Programme, United Nations publications. Paris, France.

UNEP (United Nations Environment Programme) (1996) Life cycle assessment: what it is and how to do it?, First Edition, United Nations publications. Paris, France.

UNEP/IE (United Nations Environment Programme, Industry and Environment) (1996) Cleaner production in leather tanning, a workbook for trainers. First Edition, United Nations, pp IV3-IV23.

UNEP (United Nations Environment Program) (1999) International Cleaner Production Information Clearinghouse, CD Version 1.0, Paris, France

Chapter 6

A Strategy for Sustainable Water Management in the West Bank, Palestine

A Strategy for Sustainable Water Management in the West Bank, Palestine

Abstract
Water in the West Bank in Palestine is a key issue due to its limited availability resulting from arid climate conditions as well as the Israeli control over the water resources available in the area. Since 1967, when Israel occupied the West Bank, Israel severely restricts the amount of water for the Palestinians. For example, the Palestinians have no access to the water from the River Jordan. As a consequence, there is unequal distribution of the water resources in the area between Palestinians and Israelis (50 m^3/cap/year versus 300 m^3/cap/year). The Israelis are using more than 80% of the groundwater available in the West Bank and 60% of the water flow from River Jordan. However, the current water use in the area is unsustainable because aquifers are being overexploited while deterioration of the quality of the water resources will further reduce the available quantity of good quality water. The water sector in Palestine is facing problems of water scarcity, unequal distribution of water and inadequate use of the scarce water resources and sanitation systems, in addition to the population growth and economic development which is expected to increase the pressure on the scarce resource. Therefore, there is a need for a completely new approach towards water management in the area, whereby return flows are viewed as a resource and the focus is on the conservation-oriented approach of '*use, treat and reuse*'.

After considering two initial scenarios, the "do-nothing" and the "water stress", this study develops a strategy for sustainable management of water in the West Bank that can be used to guide the development of the Palestinian water sector. Analyses of the existing situation of the water sector as well as the expected availability and demand projections for the year 2025 were conducted. It was concluded that striving for equitable water rights to the existing water resources in the area is essential to satisfy the basic water needs for all Palestinians. However, until then sustainability can gradually be achieved by the staged introduction of a combination of water management alternatives in the domestic, agricultural and industrial sectors.

Key words: Palestine, strategy, sustainability, water management, West Bank.

6.1 Introduction
The demand for high quality water is increasing all over the world because of growing populations and increasing demands from the industrial and agricultural sector. With finite resources, countries are, therefore, increasingly forced to device plans for the efficient utilization of the available water. The West Bank in Palestine is, in that sense, no exception. However, the arid climate of the West Bank region as well as the Israeli control over the water available for Palestinians living in the West Bank emphasize even more the need to be careful with available water.

On the basis of 1988-2003 numbers, the total average water use in the West Bank is 123 million m^3/year. With a population 2.5 million in 2005, this amounted to 50 m^3/cap.year. Since 2004, the water availability has remained constant while the population has increased. The Oslo II Agreement of 1995, the result of extensive negotiations between Palestine and Israel focusing on developing a state of peaceful co-existence, include a paragraph stating that the West Bank will be provided with an additional 75 million m³/year of groundwater, to be provided from "new resources" (Oslo II Agreement, 1995). Knowing the fact that there are negotiations going on between Israelis and Palestinians regarding the implementation of Oslo II promise, although very slow, we can assume that the theoretical water availability for Palestinians in the West Bank is 198 million m³/year equal to 80 m³/cap/year. The latter is expected to drop to 45 m³/cap/year by the year 2025 because of population growth and industrial development (Nazer et al., 2008). These are still very low figures by any standards, including the definition of water scarcity of less than 1,000 m³/cap/year proposed by Falkenmark (1986). It can, therefore, be stated that the West Bank is in a situation of extreme water scarcity.

The scarce water resources are facing threats of water pollution due to the disposal of untreated wastewater. This pollutes water resources and further decreases water quality and, therefore, availability (Nazer et al., 2008). In most cases wastewater is discharged directly into wadis without any type of treatment increasing the environmental problems (MOPIC, 1998 b; ADA and ADC, 2007). Depletion of water resources and deterioration of water quality are key environmental challenges that require urgent action.

Although the water resources are shared between Israel and Palestine, there is unequal allocation of the water. The per capita water use, for all purposes (domestic, industrial and agricultural), of the Israelis in 2000 is 324 m³/cap/year (adapted from Adin et al., 2004) while the Palestinians water use is 50 m³/cap/year (PWA, 2004).

A shift in thinking is needed, in terms of both water and sanitation systems, to solve the problem of water scarcity and the problem of environmental deterioration. For water systems, the attitude of dealing with water as a resource with no value needs to change. Water is to be looked at as a scarce and valuable good for which there often is no substitute. Water should be conserved whenever possible and water conservation methods must be introduced at the domestic, industrial and agricultural levels. Also a new sanitation approach is needed, one focusing on the use of the least amount of water while ensuring the improvement of public and environmental health. This is possible through a sanitation system that treats the waste on site or even derives benefits from what is considered waste by reusing it or recovering some of its components.

In order to improve the water situation in the West Bank, new ways should be travelled which change the current trend of once-through water use and identify the path towards water sustainability in the West Bank in Palestine.

In 2000 the Palestinian Water Authority (PWA) formulated its National Water Plan that outlines the direction in which the Palestinian water sector is proposed to develop until the year 2020. The main goal of the National Water Plan is to achieve the equitable and sustainable management of water resources in Palestine where everyone has access to 150 liters of water/day to satisfy his/her domestic needs and where enough water is available for the development in agriculture and industry. However, the National Water Plan relies on achieving the Palestinians' water rights, through negotiations with Israel, from the existing groundwater aquifers and surface water from the River Jordan. As the political situation is complicated and the permanent status negotiations have not been accomplished and it may take ages to achieve the water rights through negotiations, the promise of the National Water Plan is unrealistic.

The objective of this paper is to develop a strategy that allows the Palestinians living in the West Bank to access an adequate volume of water per day of sufficient quality for domestic, agricultural and industrial purposes and that leads, after a transition period, to a situation of water sustainability by 2025.

This paper consists of four sections. The first provides a general description of the situation as well as the objectives of the study. Section 2 presents a background that describes the study area, the water sector in the West Bank in terms of available resources, water use, Palestinian water rights, an analysis of the National Water Plan, a prognosis of the future development of the water sector, the existing institutional and organizational structure of the water sector and a SWOT analysis of the water sector in the West Bank. Section 3 proposes a strategy to make water a sustainable resource by 2025. The strategy includes technical, institutional and financial aspects as well as the needed regulations and awareness. Finally, section 4 presents the main conclusions.

6.2 Background
6.2.1 Study area
Historical Palestine is the area situated in the western part of Asia between the Mediterranean Sea in the west and the Jordan River and the Dead Sea in the east. It is bordered by Lebanon in the north, Syria and Jordan in the east, the Mediterranean Sea in the west and Egypt and the Gulf of Aqaba in the south (Figure 6.1 A). Historical Palestine comprises the West Bank and Gaza Strip plus what is now called Israel (Figure 6.1A). This study focuses on the West Bank (Figure 6.1). The West Bank is situated on the central highlands of Palestine; the area is bordered by the River Jordan and the Dead Sea in the east and the 1948 cease-fire line in the north, west and south. The West Bank is occupied by Israelis since 1967. In spite of the establishment of the Palestinian Authority in 1993 as a result of the peace process, the Palestinians are still struggling to declare their independent state on part of the Palestinian land, West Bank and Gaza Strip. The total area of the West Bank is 5,800 km^2 including the area of the Dead Sea that falls within the West Bank boundaries (Figure 6.1). The results of the 2007 census show that in 2007 the total Palestinian population living in the West Bank was 2.4 million (PCBS, 2008). The population growth predictions on the basis of the 2007 census are not available yet. However, the population predictions on the basis of the 1997 census indicated that the projected population of the West Bank in 2025 is 4.4 million (PCBS, 1999).

6.2.2 Available water resources

The water resources available in the area are groundwater and surface water from the River Jordan. Groundwater is the main source of fresh water for Palestinians in the West Bank as it accounts for 95% of the total water consumption. In addition, Palestinians harvest rainwater from roofs or rock catchments and store it in cisterns, to meet part of their household needs (WRAP, 1994; Haddad, 1998; Nazer *et al.*, 2008). Table 6.1 presents the water balance in the West Bank. It receives 540 mm/year of precipitation, this equals a total incoming flow from precipitation of 2970 million m^3/year out of which 679 million m^3/year infiltrates into the groundwater aquifers.

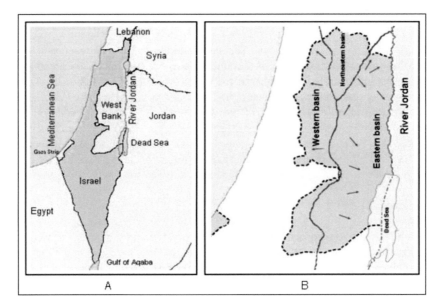

Figure 6.1 A: The West Bank Regional location, B: Ground water aquifers in the West Bank

Table 6.1: Water balance for the West Bank (Nazer *et al.*, 2008)

Item	In-flow (million m^3/year)	Out-flow (million m^3/year)
Precipitation	2970	
Runoff		77
Infiltration		679
Rainwater harvesting		7
Evapotranspiration		2207
Total	2970	2970
Note: River Jordan flow was not included in this balance because Palestinians do not have access to the water from this river		

Groundwater in the aquifer system flows in three main drainage basins: the Western, the Northeastern and the Eastern basins (Figure 6.1 B). The first two basins are shared between the West Bank and Israel, the eastern basin falls entirely within the West Bank (WRAP, 1994; MOPIC, 1998 a; SUSMAQ and PWA, 2001). The annual average recharge of the three basins is 679 million m^3/year. However, the total groundwater abstraction (fresh and brackish) by both Palestinians and Israelis amounts to 770 million m^3/year, out of which 653 million m^3/year (85%) is abstracted by Israelis and the rest (117 million m^3/year, 15%) is abstracted by Palestinians. This means that the aquifers are being overused (Nazer et al., 2008).

The River Jordan is the only source of surface water in the area. It flows from Lebanon in the north towards the Dead Sea in the south some 260 km long through a longitudinal graven known as the Rift Valley. The river flows through Syria, Lebanon, Jordan, Palestine and Israel which are its riparian states. The natural flow of the river in the absence of extraction is estimated at 1250- 1600 million m^3/year (Mimi and Sawalhi, 2003). All riparian states have the right to use part of the water from the river. However, there is no clear agreement regarding allocation of water from the river. In this context Abed and Wishahi (1999) stated that the most famous agreement regarding water allocation in this river is the Johnston Plan developed by Erick Johnston in 1953. According to this plan the water of the river is allocated as indicated in Table 6.2. The Palestinians share of water is included in the Jordanian share because the West Bank was part of Jordan at that time. Abed and Wishahi (1999) added that according to this agreement the share of Palestinians is between 257 and 320 million m^3/year. In spite of this, the Palestinians are not allowed to use water from River Jordan as a result of the Israeli control over the flow of the river.

Table 6.2: River Jordan water allocation according to the Johnston Plan in million m^3/year (Abed and Wishahi, 1999)

Lebanon	Syria	Jordan*	Israel	Total
35	132	720	400	1287
*The share of Jordan is including the share of the West Bank (257-320 million m^3/year) which at that time was part of Jordan					

6.2.3 Water use

Table 6.3 presents the water use for Palestinians in the West Bank. The overall water use has remained more or less stable, while population has grown which causes the per capita water use to decrease. The water use decreased from 139 m^3/cap/year in 1991 (Eckstein and Eckstein, 2003) to 50 m^3/cap/year in 2005 (Nazer et al., 2008). This number will continue to decrease because of the population growth and limited amount of extra water being made available for Palestinians.

Table 6.3: Overall water use for Palestinian in the West Bank

Overall (domestic +agriculture+ industrial) per capita water use					
1991 numbers [a]		1995 numbers [b]		2003 numbers [c]	
Total Mm^3/year*	Per capita (m^3/c/year)	Total Mm^3/year	Per capita (m^3/c/year)	Total Mm^3/year	Per capita (m^3/c/year)
125	**139**	139	**72**	123	**50**
* Mm^3/year stands for million m^3 per year [a] Based on Eckstein and Eckstein (2003)1991 numbers [b] Based on Fisher et al. (2005) based on 1995 numbers [c] Based on (Nazer et al. , 2008) 1988-2003 numbers					

6.2.4 Palestinian rights to water

According to Wolf (1999) water is the only scarce resource over which the international law is poorly developed and there are no internationally accepted criteria for allocating shared water resources or their benefits. However, in 1966 the International Law Association (ILA) adopted the Helsinki Rules which were further developed by the International Law Commission (ILC), an organization created by the United Nations. In 1991, the ILC completed the draft of the "Convention of the law of non-navigational uses of international watercourses" which has been approved by the United Nations General Assembly in 1997 (Wolf 1999; UN, 2005). According to article 5 of the convention, the right to use part of shared water resources by riparian states is recognized as it indicates that all parties that share an international water basin should be entitled to a reasonable and equitable share of water with the obligation of not causing significant harm to another user (article 7 of the convention). Additionally, the human right to water has been guaranteed in human rights law. The general comment # 15 released by the UN Committee on Economic, Social and Cultural Rights in November 2002 stated that "The human right to water entitles everyone to sufficient, safe, acceptable, physically accessible and affordable water for personal and domestic uses". Accordingly, both Palestinians and Israelis have the right to use the surface water coming from the River Jordan as well as the groundwater available in the area because they are sharing these water resources.

In the basin of the River Jordan, over the years many management plans and attempts to reach agreements over water resources have been proposed such as Main Plan (1953), Johnston Plan (1953), Cotton Plan (1954) and Arab Plan (1954). Nevertheless, countries in the region have continued to develop their water resources, often at the expense of other countries (Mimi and Sawalhi, 2003). For example, Israel constructed the National Water Carrier which brings some 500 million m^3/ year from Lake Teberias to the south and central region of Israel and Jordan developed the east Ghore Canal for irrigation. Historically, the Palestinians in the West Bank enjoyed using part of the water from River Jordan before the Israeli occupation of the West Bank in 1967 (Weinthal and Marei, 2001). Moreover, their right to water from river Jordan has been recognized in the Johnston Plan (1953). However, the Israelis banned the Palestinians from using the water from this river after the occupation in 1967, ignoring the Palestinians' right to the water.

Rights to surface water

In the context of surface water allocation in the area Mimi and Sawalhi (2003) developed a methodology for allocating the water from the River Jordan between the riparian parties. In their study they considered nine factors listed by the International Law Association as the factors associated with equitable water: population size, hydrology, climate, past and existing use, availability of other resources, geography of the basin, comparative costs of alternatives and economic and social needs. According to this methodology, the Palestinians would have the right to use 14% of the total flow of the river (1200-1600 million m^3/year). This amounts to some 200 million m^3 /year.

Rights to groundwater,

With regard to ground water, Eckstein and Eckstein (2003) stated that the international law constitutes the rules and norms by which states conduct their actions in relation to other states. Historically, the focus of international water law was on surface water such as shared lakes and rivers. The use, management and conservation of shared groundwater resources received little attention in international legal discourse and political circles and generally was absent from bilateral and multilateral agreements. The International Law Associations' Helsinki Rules of 1966 and Seul Rules in 1986 were among the few international documents to directly address the status of groundwater under international law (Eckstein and Eckstein, 2003). However, in the case of groundwater shared by Palestine and Israel, there is no agreement that shows the share of each party or even guarantees the water rights for both parties. Mimi and Aliewi (2005) proposed a methodology for shared groundwater allocation between Palestinians and Israelis similar to that developed by Mimi and Sawalhi (2003) for distributing water of the River Jordan. The results of their study suggested that the Palestinians have the right to 60% and the Israelis 40% of the groundwater available in the West Bank aquifers. This would give the Palestinians the right to use some 400 million m^3 /year of the groundwater available in the area.

Accordingly, if we follow Mimi and Sawalhi (2003) and Mimi and Aliewi (2005), some 600 million m^3/year of water would be available for the Palestinians in the West Bank from both the River Jordan and groundwater, a number that could satisfy the needs of the Palestinians until 2025 and beyond. However, in the existing political situation of Israel's continued denial of the Palestinian water rights, water scarcity will not decrease but, in contrast, will continue to increase.

Many ways have been proposed to confront the increasing water scarcity in the area such as desalination plants, the construction of the Red and Dead Seas canal and the peace pipeline from Turkey (Eckstein and Eckstein, 2003). All these proposals focus on finding new water resources to fill in the gap between water demand and water availability. However, re-allocation of water resources between both parties, based on equitable water use for all, is an important step towards approaching water scarcity in the area. Moreover, water management is of great importance for both Palestinians and Israelis. The focus of this paper is to reduce water demand to match water availability. The demand management methods suggested in this paper can be applied in both Israel and Palestine and will result in a decrease of the pressure on the scarce water resources. However, this paper will focus on the West Bank only.

6.2.5 Scenarios building

To determine how the water sector is going to develop over the period to 2025, two scenarios are considered. In the first, "do-nothing" scenario, it is assumed that the current political situation will, relative to the water availability issue, remain effectively unchanged. Accordingly, the water available for Palestinian use will remain under the Israeli control and will be limited to the present 123 million m^3/year. It is also assumed that the current trends in the water demand will encounter no change leaving the overall (domestic, industrial and agriculture) per capita water demand at around 50 m^3/cap/year. With the increasing population, this will mean an increasing gap between availability and expected demand which will reach some 100 million m^3/year in 2025 under the do-nothing scenario (Figure 6.2).

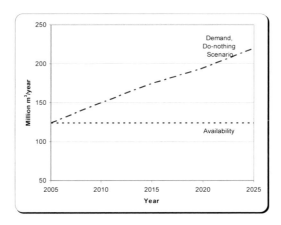

Figure 6.2: Water availability and expected demand in the "do-nothing" scenario

In the second "water stress" scenario, it is assumed that there will be some progress in the negotiations with Israelis, as a result of which the water availability will rise to 198 million m^3/year, being the existing water consumption of 123 million m^3/year plus 75 million m^3/year as agreed upon in the Oslo II agreement (1995). However, due to political, technical and financial reasons this quantity can not be made available directly so the availability will increase by some 4 million m^3 each year from the existing 123 million m^3 to 198 million m^3 by 2025. It is also assumed that the West Bank will undergo some development and improvements in the social, commercial, industrial and environmental sectors which will increase the water use. The development in the water sector will be geared towards increasing the domestic per capita water consumption to 150 liter/cap/day (this number is the target of the PWA in its national plan) to all users by 2025 (PWA,2000). For the sake of calculations, it is assumed that the population will be provided by 70 liter/cap/day in 2010 and this will increase by 5 liter/cap/day each year until 150 liter/cap/day in 2025 (Table 6.4). The water allocated for industry is taken as percentage of the domestic water use and the expected water demand for agriculture is assumed constant as the existing per capita agricultural consumption of 30 m^3/cap/year. On the basis of these assumptions, the total water demand will reach 430 million m^3/year by 2025 (Table 6.4, Figure 6.3). This will increase the gap between availability and expected demand in scenario 2 to some 230 million m^3/year (Figure 6.3).

Table 6.4: Expected domestic, industrial and agricultural water demand in the West Bank in the "water stress" scenario

Year	Population [a] million	Domestic [b] million m³/year	Industrial [c] million m³/year	Agriculture [d] million m³/year	Total million m³/year
2010	3.0	77	8	100	185
2015	3.5	121	15	116	255
2020	3.9	171	26	130	332
2025	4.4	241	43	146	430

[a] Population according to (PCBS, 1999)
[b] Domestic water demand was calculated on the basis of the assumption that the population will receive 70 liter/cap/day in 2010, this will increase by 5 liter/cap/day annually to reach 150 liter/cap/day by 2025
[c] Industrial water demand was calculated as 10%,12%,15% and 18% of domestic demand in 2010, 2015, 2020,2025, the % was based on (PWA, 2000)
[d] Agricultural water demand was calculated on the basis of assuming a constant per capita agricultural water consumption from year 2005 to 2025 (33 m³/cap/year)

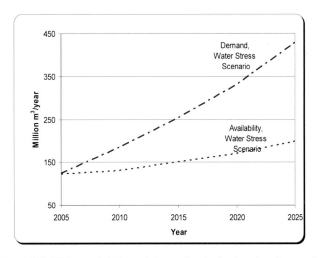

Figure 6.3: Water availability and demand under "water stress" scenario

6.2.6 Existing institutional framework of the water sector in the West Bank

The water sector in the West Bank has been under the authority of the Israeli military forces since 1967. Military orders granted full control to an Israeli Water Officer and established a permit system for drilling new wells and fixing pumping quotas. The Israeli Water Company Mekorot was given an important role in planning, implementing and operating water projects in the West Bank to the degree that Mekorot was drilling wells in Palestinian land to sell water to Palestinians. Mekorot was awarded a concession contract over water works by the Israeli Civil Administration (WRAP, 1994). Mekorot was and still is responsible for operating and maintaining wells and booster stations in Palestine; these facilities are used to supply drinking water to the West Bank Water Department (WBWD). The WBWD was formed during the Jordanian administration of the West Bank, but came under Israeli control after 1967. Since then, WBWD is responsible for the operation and maintenance of all bulk water distribution systems and, hence, for the trans-regional distribution of water resources (ADA and ADC, 2007). There has been no Palestinian institution with the responsibility for water resources management or with any powers or mandate.

The regional institutions responsible for water supply and sanitation for domestic and industrial purposes are:

Municipal water departments: these departments are found in the municipalities such as in Hebron, Nablus. Since 1967, the role of these departments is limited to the operation and maintenance of water and sewerage networks (Haddad, 1998).

Independent utilities: the Jerusalem Water Undertaking (JWU), a Palestinian company, serves the Ramallah district including the cities of Ramallah, El-Bireh, part of Jerusalem and the surrounding villages. In Bethlehem, the Water Supply and Sewerage Authority (WSSA) serves Bethlehem, Beit-Jala, Beit-Sahour and a number of villages. These Palestinian water utilities are administratively and financially independent and each has its own board of directors (Haddad, 1998).

Local committees and village councils: these committees and councils manage and develop public services in the village or locality including the supply of domestic water and sanitary services. The councils and the committees are generally unqualified from a technical, administrative and financial viewpoint, resulting in overall inefficient management (Haddad, 1998).

These regional institutions may have their own wells, mostly drilled before 1967 and/or purchase water from Mekorot. The agricultural water supply is operated by individual farmers, family farmers and collective or cooperative management associations such as the cooperative for the Fara'a water project in the Jordan valley (Haddad, 1998).

During the Israeli occupation, the water sector was facing the problems of weak institutional structure, deteriorating infrastructure and the absence of effective management of the water resources. This has been recognized in a workshop held in Berzeit University in 1994 (WRAP, 1994). In this workshop, the participants recommended a future institutional framework which consists of a Palestinian Water Authority that would have as its objective the management and efficient allocation of water to achieve the social, economic and environmental goals, a National Bulk Water Utility that would be an executive independent body with the objective to provide large-scale water services, a set of Regional Water Utilities which would be responsible to deliver services to customers and a Central Support Service Company that would provide support to Regional Water Utilities in areas which can be more efficiently provided by a central organization. One of the major conclusions of the workshop was to institutionally separate responsibilities of functions according to regulatory and service provision activities (WRAP, 1994).

After the signing of the Oslo II Agreement of 1995, it was the time for improvement and change. The Oslo II Agreement states that the Palestinian side shall assume full powers and responsibilities in the water sector except for issues that will be negotiated in the permanent status negotiations. Under the Oslo II Agreement both Israeli and Palestinian sides agreed to coordinate the management of water and sewage resources and systems in the West Bank. As a result the National Water Council (NWC) and The Palestinian Water Authority (PWA) were established in 1995. The NWC is the policy making body and is chaired by the President of the Palestinian National Authority (PNA). It consists of 11 members; these are representatives from Ministries of Agriculture, Health, Industry, Justice, Local Government, Planning and International Co-operation, Finance, representative of the municipalities, representative of water associations and unions, representative of factories and companies responsible for the distribution of water and sewerage and representative from universities. The head of the Palestinian Water Authority (PWA) serves as the Secretary General of the council as well as the liaison officer between NWC and PWA (Haddad, 1998; PWA, 2000; ADA and ADC, 2007). According to the Water Law of 2002, the National Water Council is the highest level of the water sector institutional framework. It is responsible for setting the general water policy, approval of water use plans and programs, including tariff policy, confirmation of the allocation of funds for water sector investments, and approval of the work and activities of the PWA and its annual budget. The Water Law of 2002 defines the role of PWA as the main regulatory body for water resources management and development in both West Bank and Gaza and the main official body for water supply activities such as allocating of water for beneficial uses, issuing licenses and permits for the uses of water resources, and planning and management of water supply systems (Haddad, 1998; ADA and ADC, 2007).

The Joint Water Committee (JWC) is a bilateral committee established by the Oslo II Agreement to manage water resources in the West Bank and Gaza during the temporary period until the permanent status of the peace process is settled. Permits for water and wastewater infrastructure projects must be formally issued by the Joint Water Committee (JWC). The JWC mechanism is not relevant to the current needs for the water sector development in the area. The bilateral nature of permitting water projects essentially gives power to Israel concerning any water project proposed in the West Bank. The JWC process is disruptive in nature to a sound planning of the water resources in the West Bank, as water resources management plans cannot effectively be implemented by the Palestinians because of restrictions imposed by the Israelis for accessing critical well fields. In general, the JWC has not proved to be a helpful unit in promoting fair water allocation to Palestinians, and has a history of denying Palestinian requests for changes to infrastructure (ADA and ADC, 2007).

As a result of the ongoing political situation facing the area and the halt of the peace process, the main difficulties facing the water sector are as follows:
1. The permanent status negotiations have not been accomplished as stated in the Oslo II Agreement. Therefore, the responsibilities with regard to the water sector have not been defined or transferred.

2. Although the West Bank Water Department (WBWD) acts as an executing organization of the Palestinian Water Authority (PWA), it is still supervised and controlled by the Mekorot Water Company (ADA and ADC, 2007).

3. The National Water Council is still in a transition stage and not yet properly functional. In practice; much of the work of drafting national water policy is carried out by the PWA.

6.2.7 Palestinian Water Authority's (PWA's) management plan

In 2000 the Palestinian Water Authority (PWA) developed a National Water Plan (PWA, 2000). The vision of the National Water Plan is "to ensure equitable use, sustainable management and development of Palestine's water resources." The main goals of the plan are:

1- to optimally manage, protect and conserve existing water resources and enhance new resources to meet present and future demands.

2- to guarantee the right of access to water of good quality for both the present population and future generations at costs that they can afford.

Figure 6.4 shows the overall strategy proposed by PWA in its National Water Plan (PWA, 2000). The strategy consists of three main areas where action is proposed.

Table 6.5 presents the specific objectives of the National Water Plan (PWA, 2000) and the progress achieved so far with regard to these objectives. It is important to mention that obtaining the approval for projects related to water and wastewater infrastructure from the Joint Water Committee (JWC) is difficult and time consuming because the Israelis in the JWC are positioned with veto power over any initiative or request and the bilateral nature of permitting water projects through the JWC gives Israel power to approve or reject any project (ADA and ADC, 2007).

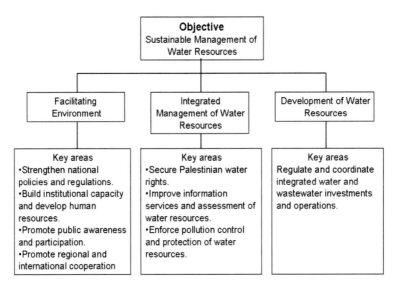

Figure 6.4: Main principles of the overall strategy of National Water Plan (PWA, 2000)

Table 6.5: Specific objectives and achievements in each objective

Objective (PWA, 2000)	Achievement (ADA and ADC, 2007)
1. Secure an equitable share of the naturally available water resources in the region.	Nothing has been achieved yet
2. Infrastructure will be expanded to progressively provide quality water service to all domestic consumers reaching an average of 150 l/cap/day by 2020. The quality of water provided to domestic consumers will meet the WHO standards.	The priority for infrastructure development in the water sector is to build new production wells and new water supply facilities especially in the areas without water services. A number of water development projects and investment programs targeting these underserved clusters have been implemented with the help of donors. These projects resulted in providing an extra 34 million m^3/year of water from groundwater through the drilling of new wells.
3. Sewer and wastewater treatment facilities will be expanded to cover 80% of the population to safeguard public health and avoid pollution. Reclaimed wastewater will be treated to standards appropriate for the relevant irrigation and for aquifer recharge.	Wastewater treatment has been neglected to a certain extent, with most attention focused to solve the water quantity and supply problems. No treatment plants have been constructed either because some of them have not been approved yet or because of lack of finance. As for pollution control and protection of water resources, the PWA has released the water law in 2002 which contains the related regulations. However, these regulations have not been enforced due to the existing political situation which weakens the Palestinian Authority in general and the Palestinian Water Authority as a result.
4. Conservation measures such as metering, leakage reduction and improved agriculture technologies will be implemented to save water.	Since the establishment of the PWA, more than 50 old networks have been rehabilitated and work is in progress to rehabilitate many more in order to reduce leakage.
5. Improve the efficiency of water services through restructuring the water sector by replacing the fragmented management and distribution system with regional bulk water utilities under the control of a single National Water Utility and by standardizing appropriate pricing system to improve cost recovery.	The establishment of the National Water Council (NWC) and the Palestinian Water Authority (PWA) is a step in the right direction towards improving the institutional structure of the water sector.

With regard to financial requirements of these projects, the NWC is supposed to support the PWA in arranging water projects funding and to be the appropriate body to establish the priority criteria of public investment programs. To date, little to no input has been made by the NWC in this critical function of investment planning, and most of the effort in arranging project financing and preparation of investment plans has been done by the PWA, with the support of donors (ADA and ADC, 2007). The Ministry of Planning coordinates the arrangement of donors' funding and proposals. However, due to the lack of financial resources, the Palestinian Authority is not supporting the investment programs. All the investments for the development of public infrastructure are funded by grants of foreign donors. Some municipalities request support of donors to finance the development of their infrastructure. They can arrange funding their development programs through bilateral arrangement with specific donors. Most of these municipal funding arrangements are not necessarily established in co-ordination with the PWA and through the Ministry of Planning but directly between the Ministry of Local Government and the donors (ADA and ADC, 2007).

The organization of the water sector presented in its National Water Plan provides clear separation between regulatory and operational (water delivery) functions and emphasize that the PWA is the sector regulator and controller of the nation's water resources. The National Water Utility proposed to be established would be responsible for overall operational tasks, including service delivery. The regional water utilities in the governorates that are proposed to be established would be responsible for delivering water to all customers (Figure 6.5). The function of National Water Utility in the West Bank is currently being undertaken by the West Bank Water Department (WBWD) (ADA and ADC, 2007).

The National Water Plan identified the role of the ministries in the effective functioning of the sector. The ministries involved in the water sector are the Ministries of Agriculture, Health, Industry, Justice, Local Government, Planning and International Co-operation and Finance.

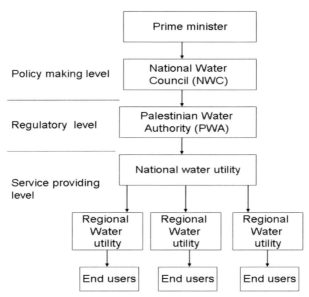

Figure 6.5: Organizational structure as presented in the National Water Plan (ADA and ADC, 2007)

6.2.8 SWOT analysis of the water sector in the West Bank

In general, a strategy builds on the existing situation, strengthening the strong elements and addressing the existing weaknesses while taking advantage of opportunities and addressing potential threats. SWOT, sometimes referred to as TOWS (Weihrich, 1982 and 1999), stands for Strengths, Weaknesses, Opportunities and Threats. SWOT analysis is a tool that helps in formulating strategic alternatives from situation analysis (Weihrich, 1982 and 1999).   According to Koo and Koo (2008), the origin of SWOT was SOFT (Satisfactory, Opportunity, Fault and Threat) which has been developed by Stanford Research Institute in the 1960s and then the F was replaced by W and it was called SWOT. Subsequently Weinhrich (1982) modified the SWOT (or what he called TOWS) into a matrix format that matches the internal factors (strengths and weaknesses) and the external factors (opportunities and threats).

The TOWS provides a framework for developing alternative strategies by analyzing strengths, weaknesses and integrating them with opportunities and threats. It provides four sets of strategic alternatives (Weihrich, 1982 and 1999). The most favorable alternative is using strengths to take advantage of opportunities (SO), another alternative is to use strengths in order of overcome threats (ST), the third is to overcome weaknesses in order to take advantage of opportunities in an attempt to turn weaknesses into strengths (WO), the fourth one, which is the least favorable situation, is facing threats in light of existing weaknesses and trying to minimize both (WT) (Weihrich, 1982 and 1999).

In this study, the TOWS Matrix was used to analyze the existing situation of the water sector in the West Bank in order to develop a strategy for the water sector. First a list of strengths, weaknesses, opportunities and threats were given after which the TOWS Matrix was developed (Figure 6.6).

The water sector has the following strengths:
S1.   Palestinians have since long adapted to water scarcity as a fact of life by using alternative water resources such as rainwater harvesting systems, rain-fed crops and virtual water.
S2.   The establishment of the National Water Council (NWC) and the   Palestinian Water Authority (PWA) is a step in the right direction towards improving the institutional structure of the water sector.
S3.   The announcement of the water law in 2002 is a manifestation of strong political commitment to improving the water sector.
S4.   There is awareness of water scarcity in government and society. However, this awareness is not sufficient to encourage the government as well as the citizens to take measures for adequately dealing with water scarcity.
S5.   The existence of NGOs that deal with water issues with qualified staff.
S6.   The Palestinian Water Authority has set the ambitious target of providing 150 l/cap/day for all domestic users by 2020.

The water sector has the following weaknesses:

W1.   The water allocated for the West Bank is not sufficient to meet the basic   needs of the growing population which is a major constraint to economic and social development.

W2.   The water services are insufficient and inefficient; around 69% of the Palestinian communities accounting for 87% of the population are connected to the water network (ADA and ADC, 2007). However, the existing facilities suffer from leakage and interruptions in supply. Only 46% of all communities have 100% coverage of water by the network. In many cases coverage is partial and deficits are made up by purchasing tanker water or relying on rain water collection (ADA and ADC, 2007).

W3.   The institutional status of the water sector in general is highly fragmented and inefficient (ADA and ADC, 2007).

W4.   Water prices do not reflect the cost of abstraction, treatment and transportation especially in agriculture with poor cost recovery.

W5.   The level of awareness about water scarcity and the new emerging technologies for dealing with it is insufficient.

W5.   Lack of community participation in water management.

The following opportunities exist for the water sector:

O1.   The availability of potential water management alternatives is an opportunity for Palestinians to increase water availability such as rain-water harvesting. Treated wastewater reuse, virtual water and changing cropping patterns are among many potential water management options that can be used in the West Bank where localities without water network and sanitation systems present an opportunity to benefit from using these emerging water and sanitation methods and to skip the phase of water wasting sanitation.

O2.   The willingness of donors to sponsor projects in the water sector is an opportunity for Palestinians to implement pilot projects which focus on water such water harvesting and reuse facilities.

O3.   Setting a pricing system for water that ensures cost recovery is an opportunity for the water sector to become financially sustainable.

O4.   If the Palestinians acquire their legitimate rights to the existing water resources, the needs as defined in the water stress scenario will be covered until 2025 and beyond.

O5.   The existence of community-based organizations, such as local cooperatives and charitable societies through which several segments of the community can be reached.

O6.   Several, local and global knowledge institutions are conducting extensive research in water management which can be used to improve the water sector.

O7.   There are many professional with the right qualifications and skills from whom the water sector could benefit much better.

The external environment poses the following threats:

T1.   The water resources are shared inequitably, whereby Israel takes control over these water resources through the occupation forces.

T2.   The constitutional status of the surface and groundwater rights in the area is unclear. There is no clear agreement regarding water allocation between Palestine and Israel.

T3.   Pollution of the water resources due to inadequate disposal of used-water.

T4.   The population growth is increasing the pressure on the scarce water resources.

Internal Factors External Factors	Strengths (S) **S1.** Adaptation water scarcity. **S2.** The NWC and PWA. **S3.** The water law of 2002 **S4.** The awareness on water scarcity. **S5.** The existence of NGOs **S6.** The challenge of providing 150 l/cap/day for domestic users.	Weaknesses (W) **W1.** Low water availability. **W2.** Insufficient and inefficient water services **W3.** The fragmented of institutional status. **W4.** Low cost recovery. **W5.** Insufficient. Awareness. **W6.** Lack of community participation.
Opportunities (O) **O1.** Potential water management alternatives and technologies. **O2.** Donor community. **O3.** Potential pricing systems that ensures cost recovery. **O4.** Palestinian water rights **O5.** The existence of community societies. **O6.** Research in water management. **O7.** Well qualified professionals.	**strengths and opportunities (SO)** **1.** Using S1, S 4 to benefit from O1, O2, O4, O6 to introduce new technologies for water saving. **2.** Establish regulations using S3 to support point 1 and to improve the pricing system O3. **3.** Activate NWC S2 in order to acquire water rights O4.	**weaknesses and opportunities (WO)** **1.** Using O1, O2, O4 to overcome W1, W2 by introducing emerging new technologies. **2.** Using O3 to alleviate W4 by introducing a new pricing system. **3.** Using O5 to overcome W6. **4.** Using O7 to overcome W3 and W5.
Threats (T) **T1.** The Israeli control over the water .resources. **T2.** There is no clear agreement regarding water allocation. **T3.** Pollution of the water resources. **T4.** Population growth.	**strengths and threats (ST)** **1.** Using S2 to overcome T1 and T2 by making use NWC to help in the negotiations with regard to the water issues. **2.** S3 can be used to overcome T3 by applying and enforcing the laws regarding pollution.	**weaknesses and threats (WT)** **1.** The threats T1, T2 regarding Israeli control over water can only be overcome through negotiations and pushing the world community to support this issue.

Figure 6.6: TOWS Matrix adapted to the West Bank water sector

In light of the above listed strengths, weaknesses, opportunities and threats the TOWS Matrix, adapted from Weihrich (1982 and 1999), was developed for the water sector in the West Bank. It presents the basis for developing a strategy for improving the water sector (Figure 6.6).

6.3 Preparing for the future

Although water availability is very low, the Palestinian Water Authority, in its National Water Plan, promises 150 l/cap/day for all domestic uses by 2020. This means that some 210 million m^3/year should be available by then for domestic purposes only, a number exceeding the existing water availability through the Oslo II Agreement of 1995 (198 million m^3/year). The National Water Plan's promise amounts to some

500 million m^3/year, from ground and surface water, for all purposes (domestic, industrial and agriculture) in the West Bank. This policy target would only be possible if the Palestinians succeed in claiming equitable rights from the existing water resources, the Jordan River and the aquifers. It may be questioned whether this target consumption is realistic and desirable, given the limited water resources in the area and the politically complicated situation which makes the achievement of the water rights rather difficult.

However, the Palestinians, in parallel with striving for their water rights, can travel an alternative path to confront the water scarcity and achieve sustainability. This can be done by applying the strategic alternatives proposed in Figure 6.6, that is, integrating strengths and weaknesses with opportunities and threats. Going this way requires: (a) technical, institutional improvements or changes; this could not be achieved without setting the needed laws and legislation to support these improvements or changes, (b) education for those who are going to implement or use them and (c) finance for the required investments in infrastructure. In the following, an approach is presented to reach water sustainability in the West Bank by the year 2025.

6.3.1 Technical level (balancing water availability and demand)

In order to improve the water sector and reduce the gap between water demand and availability, presented in the water stress scenario, the developments in the water sector in the West Bank should be geared towards sustainable water use. A simple solution to achieve this is not possible. What is needed is an integrated water resources management (IWRM) approach, intelligently combining available water management alternatives in the domestic, agricultural and industrial sectors by using strengths and making advantage of opportunities and minimizing weaknesses and threats. The "sustainable water use" scenario is proposed as an alternative to the do-nothing and water stress scenarios and aims to achieve water sustainability. This scenario suggests the following interventions in the domestic, agricultural and industrial sectors:

Domestic sector, the domestic water sector in the West Bank can use its strengths and take advantages of opportunities (SO) alternative in TOWS Matrix of Figure 6.6: in this approach the water sector can benefit from the fact that the Palestinians are well aware of the existing water scarcity and are already using water management alternatives such as rainwater harvesting systems in order to expand these alternatives and to take advantage of the existing new technologies for water saving available in the market and the willingness of donors to sponsor water management projects. In light of this, the sustainable water use scenario focuses on the reduction of the domestic water demand, through introducing water saving alternatives such as dry toilets, dual flush toilets, low-flow shower heads and faucet aerators and enhancing water reuse such as graywater systems. Moreover, rain water harvesting, which is already widely practiced, is an option to further increase alternative water resources that can substantially reduce the demand for groundwater, thus minimizing the weakness of insufficient water availability.

Demand management through introducing such alternatives has been used in several projects. For example, within a pilot project on ecological sanitation in Palestine implemented by the Palestinian Hydrology Group (PHG) and Swedish International Development Cooperation Agency (Sida), 30 units of dry toilets have been installed in Beni Naim village near Hebron city in the West Bank (Subuh, 2003; Winblad *et al.*, 2004). The project was rated as successful because it was socially accepted by the people who are using the toilets, a 25-30% reduction of water consumption and well operation of the system that prevents smell and fly breeding (Subuh, 2003). Winblad *et al.* (2004) described the project as well-conceived high standard project which showed that ecological sanitation could be an alternative for Palestine. Dry toilets have been implemented in several places such as Vietnam, China Mexico, Sweden and South Africa (Winblad *et al.*, 2004; Jönsson and Vinnerås, 2007; Mels *et al.*, 2007). These systems reduce water use and pollutant emission while increasing nutrient recovery. These systems also experienced some problems such as bad smell and clogging in the piping system which can be overcome by improved technology (Jönsson and Vinnerås, 2007; Mels *et al.*, 2007). Gray-water systems were used in the Netherlands (Het Groene Dak in Utrecht and Polderdrift in Arnhem); these cases have the potential to reduce water consumption up to 40% although partial system failures, due to lack of operation and maintenance by inhabitants, were recorded (Mels *et al.*, 2007). In Australia, the reuse scheme implemented in Sydney managed to provide 36,000 households with treated used-water for toilet flushing, garden watering and other uses, thereby reducing the demand for water from these households by 4.7 million m^3/year in 2007 (NSW, 2007).

The selection of the appropriate combination of these alternatives depends largely on the specific conditions of the location where they are proposed to be introduced. In the West Bank, there are different categories of housing locations with regard to the availability of water and sanitation services according to which the appropriate combination can be chosen. Some locations are enjoying good quality services in both water and sanitation. Other locations do not enjoy any type of good quality services and are neither connected to water distribution network nor to a sewage collection system. Most locations find themselves in between these two extremes.

A potential opportunity exists for increasing water availability and improving water services by introducing water and sanitation systems, in locations without these systems, while at the same time reducing overall water use (see Box 6.1). In locations with an existing water and sewer network, there exists a potential for reducing water use by introducing certain options without too much reconstruction requirements such as the installation of faucet aerators, low-flow shower heads and dual flush toilets. People living in these areas should be encouraged to install such fixtures. Expanding the already used alternative, rainwater harvesting systems, can be helpful in overcoming the low water availability and improve the water services. Leakage reduction is another, obvious priority activity.

Palestine is one of countries with extremely low per capita water availability. It, should therefore consider orienting the introduction of these water management options towards a situation in which each future property (separate houses, housing complexes or buildings) can be largely independent in terms of water and sanitation systems. i.e., each property has its own water resource (rainwater harvesting system) and its own wastewater treatment system (graywater treatment system in case of dry toilet use). However, this may not be possible in some places where the rainwater harvesting does not yield sufficient water. In such places part of the water needed can be supplemented with other sources which will be relatively easy due to the reduced demand in other areas. At the same time these areas can still reduce the consumption by using other water management options.

Box 6.1 Example of introducing water management options

According to ADA and ADC (2007), there are 211 localities accounting for 10% of the population in the West Bank that are neither connected to water distribution network nor to a sewerage collection system. At present, these localities depend on rainwater collected in cisterns at house level and tankers. It is proposed that these localities are made independent in terms of water and sanitation. The figure below presents a schematic of the water and sanitation systems proposed. With regard to the water system, these localities would continue using the rainwater harvesting cisterns available to cover parts of their needs. The rest of the locality water needs can be covered through a rainwater harvesting reservoir which can be constructed on the locality level combined with a water treatment plant and a local distribution system where needed. For improvement of in-house water use fixtures, the installation of low water saving fixtures such as faucet aerators (FA) and low-flow shower heads (LFSH) is proposed.

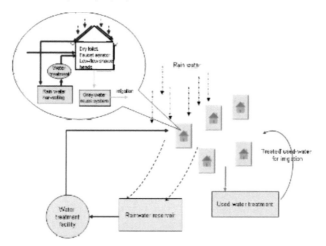

As to sanitation, a new sanitation system is proposed which is oriented towards material recovery. For toilets, there are several types of ecological sanitation systems emerging ranging from urine diversion toilets to completely dry toilets. By using dry toilets with urine diversion, the urine can be stored, collected and transferred to a facility where nitrogen and phosphorus can be recovered and used as fertilizers (Matsui et al., 2001). Solid excreta can be composted and used to produce energy; the compost can also be used as fertilizer. The collection of compost can be arranged via trucks such as those used in collecting solid waste. For other water uses in the house-hold, the remaining used-water is now graywater and can be treated on site and reused for lower water quality needs such as irrigation. This will need a local used-water collection system with a treatment plant. By doing so, less water is used, less used-water needs to be treated; and fewer pollutants need to be removed.

Agricultural sector, Agriculture is the biggest water consumer in the West Bank as it accounts for 70% of total water consumption. Therefore, effective agricultural water management is key to achieving water sustainability. The water sector can use the opportunities of water management in order to overcome the low water availability. Potential water reduction options include:

1. Treated used-water reuse: using opportunity of applying re-use options and the to overcome low water availability weakness, it is proposed that 10% of the domestic treated used-water will be reused for irrigation in 2010, 20%, 40% and 60% of the domestic treated used-water will be available for reuse in irrigation in 2015, 2020 and 2025 respectively. Although expensive in terms of investment, treated used-water reuse for agriculture has been implemented in several countries (Jordan, Tunisia and Israel), Jordan is re-using 28% of its produced used-water, Tunisia is using 13% and Israel is using 54% (Abu-Madi, 2004).

2. Expanding rain-fed agriculture: to overcome low water availability the West Bank can use the fact that agriculture in the Palestine is largely depending on rainfall, 92% of the area cultivated in the West Bank being rain-fed (PCBS, 2006); 44% of the water footprint of the Palestinians in the West Bank is from consuming rain-fed crops (Nazer et al., 2008). Therefore, it is suggested to further promote this notion by gradually reducing the area cultivated by crops that are currently irrigated but that can be cultivated under rain-fed conditions. Such crops include olives, grapes, squash, wheat and dry onion (PCBS, 2006). This approach is consistent with Falkenmark (2007), who stated that future food production will have to benefit maximally from rainfall rather than from irrigation. She maintains that climate data show that there is, also in semiarid regions, generally enough rainwater during the rainy season to meet consumptive water requirements. Yang et al. (2006) argued that rain-fed agriculture has lower opportunity costs and environmental impacts in terms of water use than irrigated agriculture. Water stored in soil moisture, or what is called green water, which is used for rain-fed crops is a free good in terms of supply. Other land uses and plants are the only major competitive users of this water. This makes rain-fed agriculture attractive in terms of cost and environmental impact although the yield of rain-fed crops is less than that of irrigated crops. With increasing water scarcity, more efforts are needed to improve rain-fed agriculture in terms of improving the yield of rain-fed crops and the efficient use of rain-fed agriculture (Yang et al., 2006; Falkenmark and Rockström, 2004).

3. Virtual water: Palestinians are already depending on virtual water through importing crops; 52% of the Palestinians water footprint is virtual water imported through in the form of crops. It is proposed that decision makers should consider increasing virtual water imports through planning the cropping patterns in a way that can reduce water use, increase the profitability of agricultural activities and ensure an acceptable level of self-sufficiency in terms of food production. That is, importing water intensive and low value crops and exporting high value crops. Many studies (Allan, J.A, 1997; WWC, 2004; Al-Weshah 2000; Hoekstra and Hung, 2005; Hoekstra and Chapagain, 2007 and Qadir et al., 2007) support the concept of virtual water trade, through food trade, as an option towards dealing with water scarcity in countries with scarce water resources. For example, Jordan has formulated policies to enable water

saving by reducing export of water intensive crops (WWC, 2004). Al-Weshah (2000) argued that although virtual water trade is a means of water saving in water scarce countries, it poses the risk of creating job loss in the agricultural sector. Planners and policy makers should consider projects to shift activities in the same area. Al-Weshah (2000) added that many voices in all the countries sharing the Jordan River are calling for better water resources management in the agriculture sector. The calls of experts from Jordan and Israel suggest that importing some agricultural products may be more rational than producing them locally in terms of their water use. One may argue that it may be unfair to call for reducing the exports of some crops from the West Bank while at the same time Israel, which is sharing the same water resources, is exporting these crops at the expense of Palestine's water share from existing water resources. According to Chapagain and Hoekstra (2004), the agriculture water use in Israel is 1290 million m^3/year out of which some 370 million m^3/year is used for producing low value wheat which can be imported from other countries. The water used to produce wheat in Israel is more than half the 600 million m^3/year proposed by Mimi and Sawalhi (2003) and Mimi and Aliewi (2005) as Palestine's water rights from the existing water resources, and is nearly double the total quantity recognized by Oslo II Agreement as the West Bank water share (198 million m^3/year).

4. The ongoing research with regard to the appropriate cropping patterns can be used to overcome the low water availability. For example Nazer *et al.* (Accepted) stated that a 4% water saving can be achieved from changing the cropping patterns according to their water use.

5. Efficient irrigation systems, there are several ways to enhance the efficiency of irrigation ranging from choosing the appropriate time for irrigation to using drip irrigation.

In the industrial sector, industry can take the opportunity of the ongoing research in re-use and recycling to overcome the weakness of low water availability and the threat of pollution. In line with this, it is proposed to enhance onsite reuse and recycling and to employ cleaner production principles in order to reduce water consumption and environmental impact. Several studies proved the feasibility of applying cleaner production principles in reducing water use in industrial activities (Carawan, 1996 a, b; Carawan and Merka, 1996; Wenzel and Knudsen, 2005; Nazer *et al.*, 2006; Fresner *et al.*, 2007). Figure 7 presents a schematic of how an industrial activity can be designed such that consumptive water use is minimized; water is provided for each process, the amount of water consumptively used (evaporated or incorporated into the product) as well as the amount and quality of return flows (used-water) from each process is to be evaluated and according to this evaluation it will be decided whether this used-water can be reused directly in another process or the type of treatment it will need in order to be reused.

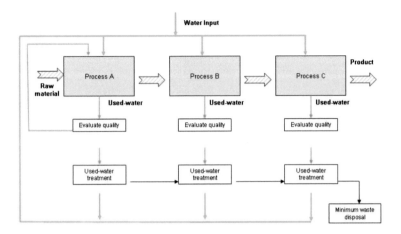

Figure 6.7: Schematic for proposed water use in an industrial activity

In line with the above strategy, the "sustainable water use" scenario was defined. In this scenario, a combination of water management alternatives was proposed in order to close the gap between availability and demand in domestic, agricultural and industrial sectors. Figure 6.8 shows that the availability of water is gradually increasing due to the use of rainwater harvesting systems while the future water demand is decreasing due to implementation of water management programs in the domestic, industrial and agricultural sectors (detailed calculations are presented in the Appendix). It can be seen that by implementing water management options the gap between availability and demand is gradually reduced through the years until it is eliminated by 2025 (Figure 6.8).

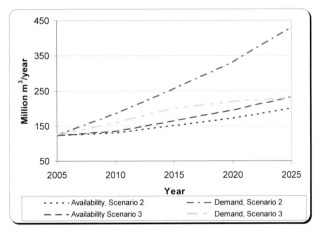

Figure 6.8: Water availability and demand under "sustainable water use" scenario

6.3.2 The needed awareness

Although there is some awareness of water scarcity, this awareness is considered insufficient because of the severe water scarcity condition facing the West Bank which needs expansion of awareness to cover all segments of the society. The West Bank can take advantage of the experience of already existing NGOs which have the potential in terms of professional human resources who can help in designing and implementing awareness campaigns. Moreover, the existing local community societies can provide logistic support for such campaigns.

There is growing recognition of the importance of social norms and attitudes in the management of water. Recent main policy documents recognize the importance of awareness raising to influence these norms and values towards a more sustainable use of water resources (Schaap and Van Steenbergen, 2001). Without changes in the way consumers use water, safe, wise and ecologically sustainable water resources management is impossible. The aim of awareness campaigns is to change the behavior of target groups towards new social norms and attitudes related to water use (Schaap and Van Steenbergen, 2001). Target groups are school children, the general public and other stakeholders. Several programs regarding water awareness have been implemented in different countries, a long list can be found in Schaap and Van Steenbergen (2001).

Education of the school children

Teaching school children the importance of water involves the inclusion of sustainable water management topics into the curriculum of primary and secondary education. Doing so provides a means of encouraging young people to understand water issues and promotes a change in their behavior towards using water (Schaap and Van Steenbergen, 2001). Studies on introducing water conservation behaviour show that the most efficient way to affect adult behavior is through educating children at school (Schaap and Van Steenbergen, 2001). The recently developed Palestinian curriculum includes text books about the environment and water issues but some of these text books are optional. These need to be made mandatory for all students and more focus should be given to the water scarcity issue and the potential methods for using water wisely. Water education in schools can be implemented by several methods such as implementing small size projects at school level regarding water issues. For example a gray-water reuse system can be implemented with the help of students in schools as a demonstration and installing water saving devices can motivate water saving. Nazer *et al.* (2007), in a study conducted in the West Bank, found evidence that involving students in water related projects has a significant effect on their awareness regarding water issues. Visits to water infrastructures facilities can also contribute to increasing awareness. Teacher training plays an important role in educating children. Teachers with the help of water managers can help to develop new ways for using local learning resources for schools and arrange workshops and seminars on water issues (GWP, 2009).

Raising the awareness of the general public and other stakeholders

Making information available to the general public is a powerful tool for creating awareness. The aim is to engage the public in such issues as water conservation and increasing the willingness to use and pay for water management alternatives. Public awareness is not a one-way communication, but an interaction of many active stakeholders, who influence each other and provide social control by mutually reinforcing agreed sets of values (GWP, 2009).

Increasing people's awareness about water scarcity and the potential water management methods in order to motivate them to use and pay for these methods can be implemented by several tools such as:

1. Using media such as radio, TV and video presentations to educate the general public about the existing water scarcity and the value of water.
2. Organizing workshops for the general public to present water management examples. In a series of workshops organized in the West Bank to assess the awareness of the general public about water management alternatives, Nazer *et al.* (2007) found that attending the workshops succeeded in improving the participants' awareness about water conservation options, including such non-traditional options as the dry toilet. This was demonstrated by the increase in the willingness to use and pay for specific water management options by the participants who attended these workshops.
3. Using newsletters, printed manuals, newspaper and electronic media reports, and email to exchange information on leading experiences and best practices in water management.
4. Organizing meetings at village level with farmers and village leaders to discuss the water scarcity issue and to present available methods to reduce the water use specially the wastewater reuse options.
5. Organizing large events such as exhibitions or festivals on special occasions.

6.3.3 The needed regulations

Educational programs and awareness campaigns are necessary but insufficient conditions to motivate everyone to use water wisely. Therefore, establishing administrative and regulatory mechanisms at appropriate levels is a complementary tool to implement policy, and, if effectively executed, can enforce the policy (Gleick *et al.*, 1995; GWP, 2009). In this context, the Water Law of 2002, considered as strength, can be used as the basis for these regulations and their enforcement. The regulations may emphasise principles to support water management, such as: polluter pays principle, public participation, ecological protection and equitable access to water resources. In the context of the West Bank the following regulations are proposed (cf. GWP, 2009).

For domestic water use:

1. Introducing water efficiency standards for in-house fixtures such as toilets, showerheads and faucets. For example, only water saving toilets are allowed to be imported. According to Gleick *et al.* (1995), these standards

      establish technological norms that ensure that a certain level of efficiency is built in of new products and services.

2. Establishing a tax system for products related to water use which enhances water saving, for example: reduce the taxes on water saving fixtures.

3. The tariff and water pricing system should reflect the cost of abstraction and distribution. Moreover, increasing the price of water will increase the people willingness to use and pay for water saving appliances. The pricing system should also take into account the affordability and protection of the poorest. According to Rogers *et al.* (2002) and Savenije and Van der Zaag (2002), the increasing block tariff system allows the poor access to water and sanitation and promote public health; under such a system poorer households get access to low-rate water since they possess fewer water consuming appliances and allows for rich-to-poor cross-subsidies and industrial-to-household subsidies (Whittington, 1997 cited in Rogers *et al.*, 2002).

4. Prohibiting the use of non-biodegradable household chemicals so as to facilitate the safe disposal and reuse of used household water.

5. Providing guidelines and standards for the used-water disposal and for the use of domestic detergents to ensure that it is free of heavy metals.

For the agricultural water use:

6. Enhancing the capacity of the agricultural extension service to advise farmers on appropriate water conservation measures.

7. Establishing guidelines and standards for the protection of groundwater through monitoring fertilizer and pesticide use, whereby farmers are trained to maintain records of quantities of key chemicals applied.

8. Ensuring that the water pricing systems for agriculture promote water saving by increasing the price of fresh water and reducing the price of treated wastewater.

For the industrial water use:

9. Establishing regulations in a way to enhance water reuse and recycling.

10. Setting-up a system for licensing of industrial firms that promote water saving.

11. Introducing the polluter pays principle by setting a reasonable charge for the disposal of wastewater into the public sewer system that ensures the reduction of pollution. Establish emission or discharge standards for acceptable concentrations of pollutants in the wastewater discharged to sewer system. Specific emission standards can be set in individual permits.

12. Establishing emission or discharge standards for concentrations of pollutants in the wastewater discharged into a sewer system. Specific emission standards can be set in individual permits.

13. Prohibiting the discharge of wastewater containing heavy metals in concentration higher than the drinking water quality standards.

6.3.4 Institutional level improvements

The main threat facing the water sector in the West Bank is the Israeli control of water resources and the absence of a clear agreement regarding water rights to the existing water resources. Given the extremely complicated political condition in the area and the continuous Israeli denial of the Palestinians role in managing the shared water resources because the lack of balance in power between Palestine and Israel, there is no choice for Palestinians to take control over water resources except through negotiations with the Israelis and putting pressure on the world community to push on Israel in order to arrive at a clear agreement that ensures equitable water shares and including control over resources.

However, the internal weaknesses of the water sector such as institutional fragmentations can be overcome by adapting the institutional structure to match the above proposed changes in the way of dealing with water. This requires a change in the institutional framework of the water sector to match the new attitude (GWP, 2000; Livingston, 2005).

In the present study, two main types of institutional structures are identified; the first type (type1) is the system that deals with the already existing water and sanitation systems and the proposed improvement in these systems while the second type (type 2) deals with the proposed new system of the water and sanitation provision.

Type 1: involves the institutional structure of communities (be these cities, villages or any other communities with the same characteristics in terms of water and sanitation systems) which are connected to a water and sanitation network. In these communities the existing technologies of water provision will continue to be used. However, more attention will be given to make people use less water through a package of different measures. In the institutional context, the existing water utilities (independent utilities such as Jerusalem Water Undertaking in Ramallah city, municipalities, village councils or local committees) or in the Regional Water Utilities as proposed in the National Water Plan of the Palestinian Water Authority, will be responsible for providing water to end users. The Palestinian Water Authority will play the role of regulator and controller of the water sector (Water law, 2002) which involves planning, licensing, allocation of water, monitoring quantity and quality and enforcement of laws and regulations. This will emphasize the separation between regulatory and delivery functions as well as emphasizing the role of the Palestinian Water Authority as the sector regulator of the nation's water resources as presented in its National Water Plan (PWA, 2000; ADA and ADC, 2007).

As for the sanitation systems, technically the used-water produced by these communities will be collected, treated and redistributed to the agriculture sector for reuse. The existing systems will be improved by constructing treatment plants where needed in order to treat all the collected used-water. The treated used-water will be conveyed for irrigation, either for irrigating parks and gardens within the same community or outside the community to farms in the surroundings of the community. Operation and maintenance of these systems will be the responsibility of existing utilities. The Palestinian Water Authority will remain responsible for the regulatory aspects of the water flows including quality monitoring and pricing.

Type 2, involves the institutional structure of communities (be these villages, small communities or emerging new housing communities or any other communities with the same characteristics in terms of water and sanitation systems) which are not connected to a water and sanitation network. On the technical level, it has been proposed earlier that such communities would be largely independent in terms of water and sanitation (Box 6.1 section 6.3.1). On the institutional level this will involve the operation and maintenance of the new in-house equipment (rainwater harvesting systems, toilets, gray-water systems…etc) which will be the responsibility of the owners. However, specialized operation and maintenance services will be made available for citizens on request at reasonable costs; this will be the responsibility of local institutions such as municipalities, village councils and community based institutions. For example, a private specialized company that can provide operation and maintenance for these equipments could be hired for the job. Monitoring of the quality of water would still be the responsibility of government (Palestinian Water Authority and Ministry of Health).

For industrial activities the water will be provided through the regional water utilities. The owners of the industries themselves will be responsible for their own water quality issues. Also effluent treatment will be the responsibilities of the industry based on strict effluent quality standards and enforcement. This will stimulate internal reuse or the sale treated used-water to irrigation farmers. However, monitoring water quality used in the context of the production of food and quality of effluent discharged to public treatment plants remains a responsibility for the government (Palestinian Water Authority and Ministry of Health).

It is important to prepare annual reports at the different institutional levels on the state of the water resources including water availability, rainfall, water storage in the aquifers, water use, water re-use, water quality and the trans-boundary water sharing issue as well as progress towards achieving sustainable water use.

Finally, a crucial component of the water sector remains the Palestinian control over water. The control of the Palestinian water resources should be transferred from the Israelis to the Palestinian side. Without getting control over these resources, the opportunities for improvement will be severely limited if not compromised.

6.3.5 The financial context
In order to evaluate the financial costs and benefits of the proposed strategy the following steps were carried out:
1. Determination of investment (one-time) costs: these relate to investment in the required infrastructure, for the acquisition of land, the construction of buildings, the purchase of equipment…etc.
2. Determination of the annual operation and maintenance costs: these relate to the cost of labor, of raw materials and energy.

3. Determination of the annual benefits or revenues from implementing the proposed strategy.
4. Determination of the present worth of the net benefits of the proposed strategy.

A cost comparison of alternatives can be done on the basis of a true financial comparison of alternatives taking into account all present and future costs. The present worth (PW) of net benefits (also called net present value NPV) is one such method which relates the cost and benefits of any activity at a certain time to the present time given certain values for discount rate (Philippatos and Sihler, 1991; Blank and Tarquin, 2005). A positive value of PW of net benefits means that the alternative is financially feasible and a negative value means that the alternative is financially infeasible. The PW of costs and benefits can be calculated according to equation (6.1).

$$PW = A\left[\frac{(1+k)^n - 1}{k(1+k)^n}\right] - I_0 \qquad \text{(Blank and Tarquin, 2005)} \qquad (6.1)$$

Where,
PW :   Present worth, is the monetary value at present or at time zero (US$).
A :   Net annual benefits (US$/year),
k :   Discount rate (year^{-1}),
n :   Number of years (year).
I_0 :   Investment in year zero (US$).

The calculations are based on the following assumptions
General assumptions
1. Calculations were based on 20 years.
2. Discount rate was estimated at 5% per year (CIA, 2009). This rate is the average discount rate of Israel and Jordan provided by CIA fact-book (2009) because a discount rate for Palestine is not available.

Domestic sector
Table 6.6 presents the costs and benefits of the proposed domestic water management interventions based on the following:
1. Cost of proposed water management options are based on costs given in table 6.
2. Cost of education has been set at 5% of investment (Hutton and Bartram, 2008) except for the rain water harvesting systems it was assumed (1%) because these are well known in the area and minimum education is needed.
3. Hutton and Bartram (2008) estimated the operation and maintenance costs for low technology options similar to those used in this strategy from 5% to 10% of investment. Given the fact that the operation and maintenance costs for rainwater harvesting systems, with the highest investment costs, are low compared to other alternatives, therefore, an estimate of 5% of investment is reasonable.

Agricultural sector

Table 6.7 presents the costs and benefits for agriculture and industrial water management interventions based on the following:

1. The calculations are based on a yearly investment for treating and distributing 5 million m^3/year, this quantity is the proposed yearly used-water reuse.

2. Costs include investment in treatment plants, collection and treatment of used-water as well as distribution systems to convey the treated used-water to farmers.

3. Costs of treatment plants vary according to type. However, the costs used in this study were adapted from Abu-Madi (2004) and were taken as average of three types of treatment most frequently used in the area, that is, activated sludge, trickling filters and lagoons. The average cost of construction is US$ 0.24/m^3, operation and maintenance is US$ 0.18/m^3 and the cost of distribution system is US$ 0.16/m^3 (Abu-Madi, 2004). Costs were adjusted by adding 10% to reflect the costs of 2009.

4. The costs of converting cropping patterns, expanding rain-fed agriculture and the virtual water trade are the costs of education and awareness programs related to them. These costs are included in the awareness costs.

5. The costs of converting existing irrigation systems into more efficient systems were not included in the calculations as these are considered the responsibility of farmers.

6. The cost of education is assumed to be 5% of investment (Hutton and Bartram, 2008).

7. Saving of fertilizers were calculated as follows: according to PCBS (2006) the West Bank spends US$ 26 million/year for fertilizers out of which 7.4% (US$ 2 million) is used for irrigated agriculture and Abu Madi (2004) stated that using treated used-water for irrigation reduces fertilizer's demand by 65%. This means that some US$ 1.3 million can be saved if all agriculture water use is replaced by treated used-water. However, the calculations are based on 5 million m^3/year out of a total water use of 83 million m^3/year. This means that some US$ 0.1 million/year can be saved.

Industrial sector:

1. The calculations are based on a yearly investment for treating 1 million m^3/year, this quantity is the proposed yearly used-water reuse. The proposed recycling and reuse implies the construction of treatment plants.

2. The investment cost of the needed treatment plants is based on Abu-Madi (2004) at a rate of US$ 0.24/m^3. Although investment costs of special treatment within industry to comply with the regulations or for reuse purposes is the responsibility of the industry itself, these costs were included by assuming double treatment costs.

Table 6.6: Potential annual costs and benefits of the proposed domestic water management interventions

Option	Rainwater harvesting system	Low-flow shower head	Faucet aerator	Leakage Prevention	Dual flush toilet	Dry toilet	Gray-water reuse system
Number of units proposed (unit/year)	25,000	30,000	100,000	50,000	20,000	10,000	10,000
Annual costs							
Investment (US$/unit/year)	4000	15	15	10	200	900	700
Life time of alternative (year)	50	10	10	10	20	20	20
Investment for 20 years	1600	30	30	20	200	900	700
Investment for 20 years	40,000,000	900,000	3,000,000	1,000,000	4,000,000	9,000,000	7,000,000
Total Investment (US$/year)				64,900,000			
Annual costs							
Operation and maintenance	2,000,000	45,000	150,000	50,000	200,000	450,000	350,000
Education	400,000	45,000	150,000	50,000	200,000	450,000	350,000
Total	2,400,000	90,000	300,000	100,000	400,000	900,000	700,000
Total all alternatives (US$/year)				4,890,000			
Annual benefits							
Quantity of water saved (m^3/unit/year)	80	7	14	8	13	28	36
water savings (US$/unit/year)	**96**	**8**	**17**	**10**	**16**	**34**	**43**
Wastewater production savings (m^3/unit/year)	0	7	14	8	13	28	36
wastewater treatment savings (US$/unit/year)	**0**	**2**	**4**	**2**	**4**	**8**	**11**
Energy savings (Kwh/unit/year)	256	28	55	32	51	109	141
Energy savings (US$/unit/year)	**18**	**2**	**4**	**2**	**4**	**8**	**10**
From using compost as fertilizer (US$/unit/year)	na	na	na	na	na	22	na
Total annual benefits (US$/unit/year)	114	12	25	14	24	72	64
Total annual benefits (US$/year)	**2,850,000**	**360,000**	**2,500,000**	**700,000**	**480,000**	**720,000**	**640,000**
Total all alternatives US$/year				8,250,000			

Notes
Investment cost of options based on (Nazer et al., submitted).
Cost of education is 5% of investment except for rainwater harvesting 1% and cost of operation and maintenance 5% of investment.
Quantity of water saved based on (Nazer et al., submitted),
Average price of water for domestic use = 1.2 US$/m^3 (PWA, 2007) and cost of wastewater treatment =0.3 US $ /m^3 (Al_Bireh, 2006)
Average price of energy = 0.07 US$/KWh (Al-Bireh, 2006)
Cost of compost is estimated on the basis of local market at $US 0.3/kg of compost, a person produce some 12kg/year (Matsui et al., 2001)
na stands for not applicable or not available

Table 6.7: Annual costs, benefits and the present worth of the proposed water management system for domestic, agricultural and industrial sectors.

Sector	Agricultural	Industrial
Investment		
Total investment for the life of the equipment (cost of equipment and installation) (million US$)	1.32	0.5
Investment for distribution system (agriculture) (million US$)	0.88	--
Investment for 20 years , I_0, (million US$)	2.2	0.5
Quantity of water saved (million m³/year)	5.0	1.0
Annual operational benefits (avoided costs)		
From water savings (million US$/year)	1.3	1.2
From wastewater treatment savings (million US$/year)	--	0.0
From energy savings (million US$/year)	--	0.3
Fertilizer savings (million US$/year)	0.1	--
Total benefits (million US$/year)	1.4	1.5
Annual operational costs		
Operation and maintenance cost (million US$/year)	0.9	0.36
Cost of treated used-water (million US$/year) purchased	0.5	0.06
Cost of wastewater discharge (million US$/m³)	--	0.06
Cost of education (million US$/year)	0.1	0.03
Total costs (million US$/year)	1.5	0.51

Notes
Investment costs for treatment plants = 0.24 US$/m³ (Abu-Madi, 2004).
Cost of distribution system is US$0.16/m³ (Abu-Madi, 2004)
Average price of water for industrial use = 1.2 US$/m³, cost of water for agriculture use = 0.3 US$/m³ (PWA, 2007)
Average cost of wastewater treatment =0.3 US $ /m³ (Al_Bireh, 2006)
Average cost of energy for abstraction of water and distribution = 0.27 US$/m³ (JWU, 2007)
Operation and maintenance for agriculture and industry 0.18US$/m³ (Abu-Madi, 2004)
Cost of treated used-water = 0.1 US$/m³ (Abu-Madi, 2004).
Cost of wastewater discharge = 0.1US$/m³
Cost of education 5% of investment (Hutton and Bartram, 2008)

The present worth (PW) of the proposed improvement in the industrial sector is positive which means that this proposed improvement is financially feasible (Table 8). The PW of the proposed improvement in the domestic and agricultural sectors is negative; this means that it is financially infeasible. This is the result of the high investment needed for these interventions and in particular the rainwater harvesting systems as it accounts for some 62% of the total yearly investment. These systems will be constructed at the expense of the users who will be willing to construct them to solve the problem of water shortage. In this context, Nazer *et al.* (submitted), in a series of workshops conducted in the West Bank, found that these systems are considered as an alternative water resource to ensure water security when the piped water supply is cut off. Many participants suffered from loosing access to water especially in summer when water is cut off for long periods. 85% of the investigated participants confirmed their willingness to pay for these systems regardless of the high cost of these systems. Besides there exist an opportunity for the Palestinians to take advantage from the donors community to provide part of the needed investment.

Table 6.8: Annual costs, benefits and the present worth of the proposed water management system for domestic, agricultural and industrial sectors.

Sector	Domestic	Agricultural	Industrial
Investment			
Total investment for the life of the equipment (cost of equipment and installation) (million US$)	116	1.32	0.5
Investment for distribution system (agriculture) (million US$)	--	0.88	--
Investment for 20 years , I_0 , (million US$)	65	2.2	0.5
Annual operational benefits (avoided costs)			
Total benefits (million US$/year)	8.3	1.4	1.5
Annual operational costs			
Total costs (million US$/year)	4.9	1.5	0.51
Net annual benefits (million US$/year) Total benefits –Total costs	3.4	-0.1	1.0
Present worth PW *	-40.2	-4.0	17.5

PW is calculated according to the equation

$$PW = A\left[\frac{(1+k)^n - 1}{k(1+k)^n}\right] - I_0 \;^{**} \text{ (Blank and Tarquin, 2005)}$$

** k is discount rate = 5 %

The calculations were based on the existing water prices which are expected to rise because of increasing water scarcity. Moreover, the social benefits gained from improved health and the gain in productive time resulted from improved health, time saving associated with better access to water and sanitation and economic gains associated with saved lives were not included in the calculations; Hutton *et al.* (2007) estimated the average rate of return of these benefits at a global average of US$ 8.1 per US$ 1 investment for combined water supply and sanitation.

For agriculture, the negative value of PW can be explained by the high investment cost of treatment plants, collection and distribution systems. However, the analyses were based on the existing low price of freshwater for agricultural purposes (US$ 0.3/m^3) compared to that for domestic and industrial purposes (US$ 1.2/m^3). Increasing the price of freshwater by US$ 0.1/m^3 will increase the PW from minus 4.0 million US$/year to 8.6 million US$/year. Maintaining a reasonable price for freshwater used for irrigation can help in making the implementation of water management strategies financially feasible. It also can encourage farmers to use treated used-water for irrigation.

Water is to be looked at as a scarce and valuable good for which there often is no substitute. This statement is reminiscent to, but subtly differs from one of the principles of the International Conference on Water and Environment in Dublin 1992, namely that "Water has an economic value in all its competing uses and should be recognized as an economic good as well as a social good." Managing water as a scarce and valuable good is an important way of achieving social objectives such as efficient and equitable use and encouraging conservation and protection of water resources. In this context, Rogers *et al.* (2002) argued that water pricing can play an important role in encouraging people to adopt water demand management options. Rogers *et al.* (2002) added that higher water rates allow utilities to extend services to those currently not served and price policies can help maintain the sustainability of the resource when the price reflects its true cost. The effects of price policy are demand reduction, efficient reallocation of resources and increasing supply (Rogers *et al.*, 2002; Savenije and Van der Zaag, 2002).

6.4 Conclusions
The objective of this paper was to develop a strategy for the sustainable management of the water in the West Bank. Three scenarios were discussed; the "do-nothing" scenario which assumes that the existing water availability will encounter no change due to the existing political situation which allows the Israelis to restrict the water availability while the population is increasing, thereby increasing the water demand. The "water stress" scenario assumes that the water availability will increase due to improved negotiations between Palestinians and Israelis; however, population growth and the development and improvements in the social, commercial, industrial and environmental sectors will increase the demand for water. The "sustainable water use" scenario proposes a strategy for the sustainable water management.

Within the limitations of the study on making water use in the West Bank sustainable, the following conclusions may be drawn:

1. Under both the "do-nothing" and "water stress" scenarios there is an increasing gap between water availability and water demand.

2. The gap between water availability and water demand from Palestinians in the West Bank can be closed by gradually introducing water management alternatives that increase the availability (through rain-water harvesting) and reduce the demand through water conservation options as well as re-use options.

3. The proposed alternatives in the industrial sectors proved to be financially feasible on the basis of the existing water price.

4. In the domestic sector the proposed methods were financially infeasible because of the high investment required for the new interventions. However, the social benefits gained from improved health and social life, not included in the calculations, may justify these investments

5. In the agricultural sector the proposed methods were financially infeasible because of existing low prices for agriculture. Increasing water prices in the agricultural sector will motivate farmers to use treated used-water for irrigation.

6. Legislation and regulations regarding the introduction of these alternatives is an important supporting tool. Awareness and education about water scarcity and potential methods for dealing with it is crucial to achieve effective management.

References

Abed, A. and Wishahi, S. KH. (1999) Geology of Palestine West Bank and Gaza Strip, Palestinian Hydrology Group (PHG), Jerusalem. In Arabic.

Abu-Madi, M. (2004) Incentive Systems for Wastewater Treatment and Reuse in Irrigated Agriculture in the MENA Region: Evidence from Jordan and Tunisia, PhD Dissertation, Delft University of Technology and UNESCO-IHE Institute for Water Education, Taylor and Francis Group plc, London, UK.

Adin, A., Netanyahu, S., Tzvilevinson, (2004), Israeli-Palestinian Water Issues: Israeli perspective, Report presented to the International Committee of the Red Cross, Israel.

Al-Weshah, R.A. (2000) Optimal use of Irrigation Water in the Jordan Valley: A Case Study, Water Resources Management Journal, volume 14, pp 327-338.

Allan, J.A.(1997) "Virtual water": A long term solution for water short Middle Eastern economies? Water Issues Group, School of Oriental and African Studies (SOAS), University of London, Paper presented at the 1997 British Association Festival of Science, Water and Development Session, 9 September 1997.

Austrian Development Agency (ADA) and Austrian Development Cooperation (ADC) (2007) Water Sector Review, West Bank and Gaza, Volume I- summary Report (Final Report), prepared by Jansen and Consulting Team water consultant to the Austrian Development Agency for Palestine/Israel/Jordan, Jerusalem.

Blank, L. and Tarquin, A. (2005) Engineering Economy, 6th ed, International edition, McGrawHill.

Carawan, R.E.(1996 A) Liquid Assets for Your Bakery, North Carolina Cooperative Extension Service. Accessed online sep. 2008,
http://www.bae.ncsu.edu/bae/programs/extension/publicat/wqwm/food.html.

Carawan, R.E.(1996 B) Liquid Assets for your Diary Plant, North Carolina Cooperative Extension Service. Accessed online sep. 2008,
http://www.bae.ncsu.edu/bae/programs/extension/publicat/wqwm/food.html.

Carawan, R.E. and Merka, B. (1996) Liquid Assets for Your Poultry Plant, North Carolina Cooperative Extension Service. Accessed online sep. 2008,
http://www.bae.ncsu.edu/bae/programs/extension/publicat/wqwm/food.html

Chapagain, A. K and Hoekstra, A. Y. (2004), Water footprint of Nations, Value of Water Research Report series No. 16 volume 1 and 2, UNESCO-IHE institute for Water   Education,   Delft-The   Netherlands.
http://www.waterfootprint.org/Reports/Report16.pdf., accessed August 2006

CIA,   (2009)   CIA-The   world   fact   book   West   Bank.
http://www.cia.gov//library/publications/the_world_factbook/geor/.html., accessed March 2009.

Eckstein, Y. and Eckstein, G., (2003) Groundwater Resources and International Law in the Middle East Peace Process, Water International, Vol 28, pp.154-161.

Falkenmark,M.(1986), Fresh water- time for a modified approach, Ambio, 15(4): 192-200.

Falkenmark, M., and J. Rockström, (2004), Balancing water for humans and nature; the new approach in ecohydrology. Earthscan, London; 247 pp.

Falkenmark, M. (2007),Shift in Thinking to Address the 21st Century Hunger Gap, Moving Focus from Blue to Green Water Management, Water Resources Management , volume 21, pp3-18.

FAO (Food and Agriculture Organization of the United Nations), AQUASTAT (2002), Online statistics, http://www.fao.org/nr/water/aquastat. , accessed March 2009.

Fisher, F., Huber-Lee, A., Amir, I., Arlosoroff, S., Eckstein, Z., Haddadin, M., Hamati, S., Jarrar, A., Jayyousi, A., Shamir, U., Wesseling, H., (2005), Liquid Assets: An Economic approach for Water Management and Conflict Resolution in the Middle East and Beyond, Resources for the Future (RFF), Washington, DC USA.

Fresner, J., Schnitzer, H., Gwehenberger, G., Planasch, M., Brunner, C., Taferner, K., and Mair, J., (2007), Practical experiences with the implementation of the concept of zero emissions in the surface treatment industry in Austria, Journal of Cleaner Production, 15, pp. 1228-1239.

Gleick, P. H., Loh,P., Gomez, S. V. and Morrison, J., (1995) California Water 2020 : A Sustainable Vision, Pacific Institute for Studies in development, Environment and Security, Calefornia.

GWP (Global Water Partnership) (2000), Integrated Water resources management, Technical Advisory Committee (TAC) Background paper no. 4, Stockholm, Sweden.

GWP (Global Water Partnership) (2009) Global Water Partnership ToolBox, social change instruments http://www.gwptoolbox.org/ accessed January 2009

Haddad, M. (1998) Planning water supply under complex and changing political conditions: Palestine as a case study, Water Policy 1 , pp 177-192.

Hoekstra, A. Y. and Hung, P.Q (2005), Globalization of Water Resources: international water flows in relation to crop trade, Global Environmental Change, volume 15, pp 45-56.

Hoekstra, A. Y. and Chapagain, A. K (2007), Water footprint of nations: Water use by people as a function of their consumption pattern, Water Resources Management, volume (21), pp 35-48.

Hutton, G., Haller, L. and Bartram, J.(2007), Economic and health effects of increasing coverage of low cost household drinking-water supply and sanitation interventions to countries off-track to meet MDG target 10, Public Health and the Environment, World Health Organization (WHO), Geneva.

Hutton, G. and Bartram, J., (2008), Global costs of attaining the Millennium Development Goal for water supply and sanitation Bulletin of the World Health Organization, 86 (1), pp 13-19.

Jerusalem Water Undertaking (JWU) (2007) Water prices, Rates and Tariffs, Ramallah, Palestine, accessed June 2007 http://www.jwu.org.

Jönsson, H. and Vinnerås, B., (2007) Experiences and suggestions for collection systems for source-separated urine and faeces, Water Science and Technology, Advanced Sanitation, 56(5), pp71-76.

Koo, L.C. and Koo, H., (2008) Developing Strategies for the Government of Macau, SAR with SWOT analysis, Paper presented in the 3rd International Conference on "Public management in 21st Century: Opportunities and Challenges, October 14-15 2008, Macau.

Livingston, M.L., (2005), Evaluating changes in water institutions: methodological issues at the micro and meso levels, Water Policy, 7, pp. 21-34.

Matsui, S., Henze, M., Ho, G. and Otterpohl, R. (2001) Emerging Paradigms in Water Supply and Sanitation, In Frontiers in Urban Water Management: Dead Lock or Hope, IWA publishing and UNESCO, UK.

Mels, A., Van Betuw, W. and Braadbaart, O., (2007) Technology Selection and Comparative performance of source-separating Wastewater Management Systems in Sweden and the Netherlands, Water Science and Technology, Advanced Sanitation, 56(5), pp77-85.

Mimi, Z. and Aliewi, A. (2005) Management of Shared Aquifer Systems: A Case Study, The Arabian Journal for Science and Engineering, Vol.30, Number 2C. King Fahd University of petroleum and Minerals, Dhahran, Kingdom of Saudi Arabia. Available online http//:www.kfupm.edu.sa/publications/ajse/ , accessed June 2007.

Mimi, Z and Sawalhi, B., (2003) A Decision Tool for Allocating the waters of the Jordan River Basin between all Riparian Parties, Water Resources Management 17, pp 447-461.

MOPIC (Ministry of Planning and International Cooperation) (1998 a) Emergency Natural resources Protection Plan for Palestine "West Bank Governorates", Ministry of Planning and International Cooperation, Palestine.

MOPIC (Ministry of Planning and International Cooperation) (1998 b) Regional Plan for the West Bank Governorates, Water and Wastewater Existing Situation 1st ed., Ministry of Planning and International Cooperation, Palestine.

Nazer, D.W., Al-Sa'ed, R.M. and Siebel, M.A. (2006), Reducing the environmental impact of the unhairing-liming process in the leather tanning industry, Journal of Cleaner Production vol.14 # 1:pp 65-74.

Nazer, D.W., Siebel, M.A., Mimi, Z. and Van der Zaag, P. (2007), Reducing Domestic Water Consumption as a Tool to Raise Water Awareness in the West Bank, Palestine, paper presented in the 13th International Sustainable Development Research Conference, Vasteras, Sweden June 10-12 2007.

Nazer, D.W., Siebel, M.A., Mimi,Z., Van der Zaag, P. and Gijzen, H.J. (2008), Water footprint of the Palestinians in the west Bank, Palestine, Journal of American Water Resources Association (JAWRA).

Nazer, D.W., Siebel, M.A., Mimi, Z., Van der Zaag, P. and Gijzen, H. (Submitted), Financial, Environmental and Social evaluation of Water Management Options in the West Bank, Palestine. Journal Water Resources Management.

Nazer, D.W., Tilmant, A. , Siebel, M.A., Mimi,Z., Van der Zaag, P. and Gijzen, H. (Accepted), Optimizing irrigation water use in the West Bank, Palestine. Submitted to Agricultural Water Management.

Oslo II Agreement (1995), Israeli-Palestinian Interim Agreement on the West Bank and the Gaza Strip, Annex III, Article 40, Washington D.C., September 28 1995.

PCBS (Palestinian Central Bureau of Statistics) (1999) Population in the Palestinian Territory 1997-2025 , Palestinian Central Bureau of Statistics, Palestine.

PCBS (Palestinian Central Bureau of Statistics) (2006) Agricultural Statistics 2004/2005, Palestinian Central Bureau of Statistics, Palestine, available on line, http://www.pcbs.org

PCBS (Palestinian Central Bureau of Statistics) (2008), Population, Housing and Establishment Census 2007, Census Final Results in the West Bank, Summary (Population and Housing), Ramallah- Palestine. available on line, http://www.pcbs.org

Philippatos, G. C., and Sihler, W.W. (1991) Financial Management text and cases, 2nd ed., Allyn and Bacon, Massachusetts.

PWA (Palestinian Water Authority) (2000) National Water Plan, Final copy. Palestinian Water Authority, Palestine.

PWA (Palestinian Water Authority) (2004) personal communication with engineer Yaakoub Yahya on estimated industrial water consumption, unpublished data. Palestinian Water Authority, Palestine.

PWA (Palestinian Water Authority) (2007) personal communication with engineer Adel Yasin, unpublished data. Palestinian Water Authority, Palestine.

Rogers, P., Silva, R. and Bhatia, R. (2002), Water is an Economic Good: How to use prices to promote equity, efficiency and sustainability, Water Policy, volume 4, pp 1-17.

Saleth, R.M., and Dinar, A., (2004), The Institutional Economics of Water: A Cross-Country Analysis of institutions and performance, The World Bank and Edward Elgar publishing. U.K

Savenije, H.H.G., and P. van der Zaag, 2002. Water as an economic good and demand management; paradigms with pitfalls. Water International 27(1): 98-104

Schaap, W. and Van Steenbergen, F. (2001) Ideas for Water Awareness campaigns, Global Water Partnership, Stockholm, Sweden.

Shaheen, H. (2003), Wastewater Reuse and Means to Optimize the Use of Water Resources in the West Bank, Water International, 28(2), pp. 201-208.

Subuh, Yousef. (2003) Ecological Sanitation based on Urine Diversion Technology (Dry Sanitation) in Palestine: A Pilot Project, EcoEng Newsletter, Number 8, International Ecological Engineering Society, Switzerland. http://www.iees.ch/EcoEng032/EcoEng032_Subuh.html. accessed March 2009.

SUSMAQ (Sustainable Management of the West Bank and Gaza Aquifers) and PWA (Palestinian water Authority) (2001) Data Review on the West Bank Aquifers, working report SUSMAQ-MOD #02 V2.0, version2, Water Resources Systems Laboratory, University of Newcstle Upon Tyne and Water Resources and planning Department, Palestinian Water Authority.

United Nations (UN), (2005), Convention on law of non-navigational uses of international watercourses, adopted by the General Assembly in 1997, available on line http://untreaty.un.org/ilc/texts/instruments/english/conventions/8_3_1997.pdf. Accessed January 2009.

Weihrich, H. (1982), The TOWS Matrix—a Tool for Situational Analysis, Long Range Planning,15(2), pp 54-66.

Weihrich, H. (1999), Analyzing the Competitive Advantages and Disadvantages of Germany with TOWS Matrix—An Alternative to Porter's Model. European Business Review, 99(1), pp 9-22.

Weinthal, E. and Marei, A. (2001), One Resource, Two Visions: The prospects for Israeli-Palestinian Water Cooperation, Globalization and water resources management: Changing Value of Water, AWRA/IWLRI-university of Dundee International Specialty Conference.

Wenzel, H. and Knudsen, H. (2005), Water Savings and Reuse in the Textile Industry, in Modern Tools and Methods of Water Treatment for Improving Living Standards, pp.160-189, Springer, Netherlands.

Winblad, U., Simpson-Hebert, M., Calvert, P., Morgan, P., Rosemarin, A., Sawyer, R. and Xiao, J., (2004), Ecological Sanitation: revised and enlarged edition, Stockholm Environment Institute, Stockholm, Sweden.

Wolf, A.T., (1999), Criteria for equitable allocation: the heart of international water conflict, Natural Resources Forum, 23, pp 3-30.

WRAP (1994) Palestinian Water Resources Project, A Rapid Interdisciplinary Sector Review and Issues Paper, The Task Force of the Water Resources Action Program, Palestine.

WWC (World Water Council) (2004), E- conference Synthesis , Virtual Water Trade – conscious choices. WWC publication No.2.

Yang, H., Wang, L., Abbaspour, K. And Zehnder, A. (2006) Virtual water and the need for greater attention to rain-fed agriculture, Water 21, Magazine of the International Water Association, April 2006.

Appendix
Calculations of potential reduction due to implementation of water management proposed strategy

1. Domestic sector

An example of potential reduction in water demand and percentage of households served by the proposed interventions is presented in table A-1 and was based on the following assumptions were used in the calculations:

1. 25,000 units/year of rainwater harvesting systems is to be constructed.
2. 30,000 units/year of low flow shower heads are to be installed.
3. 100,000 units/year of faucet aerators are to be installed.
4. 20,000 units/year of dual flush toilets are to be installed.
5. 10,000 units/year of dry toilets are to be installed.
6. 10,000 units/year of is gray-water reuse systems are to be installed.
7. On yearly basis a 50,000 household will take measure to reduce its water consumption by 10% (8 m^3/year) through leakage prevention; this will include all households by 2025.

Table A-1 Example of potential reduction in water consumption and percentage of households served by the proposed interventions

Year		2010		2015		2020		2025	
Population (x million)		3.0		3.5		3.9		4.4	
Number of households (x 1000)		500		583		650		733	
Proposed option	Savings per unit m³/year*	Number of units 1000	Reduction in water consumption (Million m³)	Number of units 1000	Reduction in water consumption (Million m³)	Number of units 1000	Reduction in water consumption (Million m³)	Number of units 1000	Reduction in water consumption (Million m³)
Rain water harvesting [a]	80	50	4.0	175	14.0	300	24.0	425	34.0
% of households served		26		44		58		69	
Low-flow shower heads	7	30	0.2	180	1.3	330	2.3	480	3.4
% of households served		6		31		51		65	
Faucet aerators	4.7	100	0.5	600	2.8	1100	5.2	1600	7.5
% of households served		7		34		56		73	
Leakage prevention***		50	0.4	300	2.4	550	4.4	733	5.9
% of households served		10		51		84		100	
Dual flush toilets	13	20	0.3	120	1.6	220	2.9	320	4.2
% of households served		4		21		34		44	
Dry toilets	28	10	0.3	60	1.7	110	3.1	160	4.5
% of households served		2		10		17		22	
Graywater reuse	36	10	0.4	60	2.2	110	3.9	160	5.8
% of households served		4		21		34		44	
Total			6.1		26		45.8		65.3

[a] already some 80,000 units in operation these have been included in the calculations of the % of household covered.

*Quantity of water saved is based Nazer et al. (submitted a)

** Each unit is expected to serve 1 family except faucet aerators where each family needs 3 of them.

*** Leakage prevention reduces the house-hold consumption by 8m³/house-hold , it is assumed that 10% of the house-holds will be concerned about leakage prevention

2. Agricultural sector: Table A-2 presents the potential water reduction due to implementation of the water management alternatives. Calculations were based on the following assumptions.

1. Treated used-water reuse:
As a perquisite for reusing treated used-water (usually known as wastewater) in agriculture is to make sure that the used-water is free of toxic chemicals, this can be achieved through restrictions of domestic chemicals.

- The quantity of treated used-water reuse is calculated by assuming that the used-water production is 80% (Haddad, 1993 cited in Shaheen, 2003) of the domestic water use.
- 10% of the domestic treated used-water will be reused for irrigation in 2010, 20%, 40% and 70% of the domestic treated used-water will be available for reuse in irrigation in 2015, 2020 and 2025 respectively.
- The domestic water use was assumed to be the water demand in table 5 minus the reduction in demand due to implementing water management options presented in table A-1. for example in 2010 the demand is 77 million m^3 and the reduction in the demand is 5.8 million m^3, so water-use is 71.2 million m^3 and used-water production is 71.2 X 0.8= 57 million m^3, 10% of this is assumed to be used which equals 5.7 million m^3 approximately 6 million m^3.

2. Expanding rain-fed agriculture: irrigation of crops which are feasible under rain-fed irrigation system will be reduced gradually until it is eliminated by 2025. Crops included are olives, grapes, squashes, wheat and dry onion. Accordingly, 20% of the area will be reduced by 2010; the percentage will increase to 40%, 70% and 100% in the years 2015, 2020 and 2025 respectively. The irrigated area of the above crops was taken from (PCBS, 2006) and the water requirement was calculated according to the methodology explained in Nazer *et al.* (2008).

3. Virtual water: 2% reduction was assumed due to the reduction of the export of water intensive crops such as citrus and bananas.

4. Change in cropping patterns: a 4% water saving can be achieved from changing the cropping patterns (Nazer *et al.*, Submitted c).

5. Efficient irrigation systems: 5% reduction in water use was assumed because it is difficult to exactly calculate how much water can be reduced due to the use of these types of irrigation systems and because the saved quantity of water varies significantly from type to type.

6. Leakage prevention was assumed to be reduced by 5%, 10% and 15% in 2015, 2020 and 2025 respectively.

Table A-2 Water demand reduction due to implementation of water management in the agricultural sector

Year	Quantity of water demand reduced 10^6 m³/year			
	2010	2015	2020	2025
Existing demand	100	116	130	146
Proposed option				
Treated used-water reuse	6	15	40	96
Change of crop patterns (4%)	4	5	5	6
Expanding rain-fed agriculture	2	4	7	10
Virtual water trade (2%)	2	2	3	3
Efficient irrigation (10%)	10	12	14	16
Leakage prevention (% of demand)	0	6	13	22
Total reduction in demand	**24**	**46**	**82**	**153**
Percentage reduction in demand	**24**	**40**	**63**	**105**

3. Industrial sector, The expected demand reduction due to implementing the proposed water management scheme in the West Bank is presented in Table A-3. There are no figures about the industrial water consumption in the West Bank; industrial consumption is usually included in the domestic water consumption. However, there exist estimates about industrial activities. These estimates were used for the calculations of the expected demand reduction in the industry (PWA, 2004).

Table A-4 is a summary of the availability and demand for water stress scenario and the sustainable water use scenario.

Table A-3 Expected water consumption reduction in the main industries due to implementation of water management programs.

Industry	Consumption (% of total)*	Expected reduction in water consumption due to reuse or recycle (%)	2010		2015		2020		2025	
			Expected Consumption* (10^6 m³/year)	Expected reduction In consumption (10^6 m³/year)	Expected Consumption (10^6 m³/year)	Expected reduction In consumption (10^6 m³/year)	Expected Consumption (10^6 m³/year)	Expected reduction In consumption (10^6 m³/year)	Expected Consumption (10^6 m³/year)	Expected reduction In consumption (10^6 m³/year)
Total industry (from table 5)	100		8.0		15.0		26.0		43.0	
Food industry	20	50 [1]	1.6	0.8	3.0	1.5	5.2	2.6	8.6	4.3
Textile	4	40 [2]	0.3	0.1	0.6	0.2	1.0	0.4	1.7	0.7
Tanning	2	58 [3]	0.2	0.1	0.3	0.2	0.5	0.3	0.9	0.5
Other industries	74	30 [4]	5.9	1.8	11.1	3.3	19.3	5.8	31.8	9.5
Total			8.0	2.8	15.0	5.2	26.0	9.1	43.0	15.1

* the percentage for water consumption in the different types of industries is based on estimates of PWA (2004)

References

[1] Reduction in water use in food industry varies from industry to industry for example in dairy a 70% reduction can be achieved (Carawan,1996 b), for bakery 50% reduction was found (Carawan,1996 a) and poultry 67% (Carawan and Merka,1996) according to these references a 50% reduction was assumed.

[2] Percentage is based on Wenzel and Knudsen (2005)

[3] Percentage is based on Nazer *et al.* (Submitted a)

Table A-4 Proposed future water demand reduction due to implementation of water management programs in the domestic, industrial and agricultural sectors.

	Year	Quantity of water saved 10^6 m^3/year			
		2010	**2015**	**2020**	**2025**
water stress scenario	Availability*	131	151	171	198
	Demand	185	255	332	430
	Gap	**54**	**104**	**161**	**232**
Sustainable water use scenario	Availability — Increase of availability due to rainwater harvesting	4	14	24	34
	Total availability (availability in water stress scenario + increase)	**135**	**165**	**195**	**232**
	Demand reduction — Reduction in domestic demand	2	12	22	31
	Reduction in agricultural sector	24	46	82	154
	Reduction in industrial sector	3	5	9	15
	Total reduction in demand	29	63	113	200
	Demand (reduced demand)	**156**	**192**	**219**	**230**
	Gap	**21**	**27**	**24**	**-2**

* water availability was calculated on the basis of the assumption that availability will increase by some 4 million m^3 each year from existing consumption 123 million m^3 to 198 million m^3 by 2025.

Chapter 7

Conclusions and Recommendations

Conclusions and recommendations

7.1 Conclusions

The aim of the research was to develop a framework for the sustainable management of water resources in the West Bank in Palestine. The research has three objectives, which are evaluated in this concluding chapter.

Objective 1: To prepare a water balance and an inventory analysis of the water use in the West Bank in Palestine by determining its water footprint.

The water footprint is defined in this study as the total volume of fresh water used to produce goods and services consumed by the individual, business or nation (consumption component) plus the volume of fresh water needed to somehow assimilate the waste produced by that individual, business or nation (contamination component).

It was found that the consumption component of the water footprint of the West Bank was 2791 million m^3/year. Approximately 52% of this is virtual water consumed in its virtual form through consuming imported goods such as crops and livestock, the production of which used water in the country of production. The rest is consumed from internal resources; these are rainwater stored in soil moisture and up taken by crops and the water withdrawn in its real sense from groundwater or surface water (blue water).

On a per capita basis the consumption component of the water footprint was 1,116 m^3/cap/year while the global average is 1,243 m^3/cap/year. Out of this number 50 m^3/cap/year was withdrawn from water resources available in the area. Only 16 m^3/cap/year (1.4%) was used for domestic purposes. This number is only 28% of the global average and 21% of the Israeli domestic water use. The West Bank is suffering from severe water scarcity. Therefore, there is a need for a shift in thinking about water management in the West Bank, a thinking that supports the approach of *'use, treat and reuse'* instead of the common approach of *'use and dispose'*.

Objective 2: To evaluate alternatives for water use, suitable for Palestinian social, cultural, religious and economic conditions, now and in the near future, so as to reach water sustainability by 2025.

This objective was accomplished by developing case studies aiming at demonstrating the feasibility of appropriate water management at the domestic, industrial and agricultural levels.

The first case study was conducted in the domestic sector. The objective of this case study was to evaluate the environmental, financial and social impacts of potential domestic water management options using Life Cycle Impact Assessment (LCIA) and to propose a water use scheme for the *house of tomorrow*, in line with the *use-treat-reuse* approach. The main findings of this study was that by introducing a combination of water management options in the domestic area, a substantial decrease in water consumption of up to 50% could be achieved, thereby reducing the pressure on the scarce water resources. This reduction was achieved because the water was used more than once and not at the expense of water use purposes; thus reducing the use of groundwater. Environmental and financial impacts can be reduced similarly; the annual environmental impact of the in-house water use can be reduced by some 8% when using low-flow shower heads, and up to 38% when using rainwater harvesting systems. Moreover, the annual financial benefits that can be achieved were found to range from US$ 51 for leakage prevention up to US$ 222 for rainwater harvesting systems. In the social context, it was found that introducing such options can improve the quality of life of those not having sufficient water. It was concluded that, theoretically, the *house of tomorrow* can be largely independent in terms of water and sanitation. It was also concluded that there is a popular willingness to take part in water conservation in the domestic sector in the West Bank. The strongest driving force for using water conservation measures is the awareness that water is a scarce resource. Education and awareness campaigns in the context of water management with a focus on non-traditional options, such as dry toilets, are key to achieving sustainable water use.

The objective of the second case study was to find the optimal cropping patterns in the West Bank in order to reduce water use for irrigation while at the same time maximizing profits using a linear programming model. It was concluded that water scarcity can be approached by changing the cropping patterns according to their water use. Moreover, it can simultaneously increase the added value per unit of water. The linear programming model showed that a reduction of 4% of the total agricultural water use in the West Bank can be achieved by changing cropping patterns. Expanding rain-fed agriculture is an effective method to address water scarcity; it can substantially reduce the agricultural national water consumption.

The objective of the industrial case was to reduce the water consumption, environmental impact and the production cost of the unhairing-liming process in the leather tanning industry as a sample water consuming industry. The conventional process of unhairing-liming uses, for each new batch of hides or skins, fresh water and chemicals according to the standard recipe and discharges the used process water. This study proposes a method for unhairing-liming of hides that uses fresh water and chemicals for the first batch of hides or skins but that subsequently reuses this water and these chemicals for subsequent batches of hides or skins. Thus reduced quantities of water and chemicals are used. It was concluded that the unhairing-liming process water could be reused four times without affecting the quality of the final leather product. The proposed unhairing-liming treatment permits savings in water consumption and wastewater production of up to 58%. The chemicals consumption was reduced; sodium sulfide by 27% and lime by 40%. Moreover, there was a reduction of the overall environmental impact of up to 16%.

> **Objective 3:** To develop a strategy for the sustainable water management in the West Bank.

Three scenarios were discussed. The "do-nothing" scenario assumes that the existing water availability will encounter no change due to the existing political situation which allows the Israelis to restrict the water availability while the population is increasing, thereby increasing the water demand. The "water stress" scenario assumes that water availability will increase due to improved negotiations between Palestinians and Israelis; however, population growth and the development and improvements in the social, commercial, industrial and environmental sectors will increase the demand for water. The "sustainable water use" scenario proposes a strategy for the sustainable water management.

It was concluded that under both the "do-nothing" and "water stress" scenarios there is an increasing gap between water availability and water demand. The "sustainable water use" scenario shows that this gap can be closed by gradually introducing water management alternatives that increase the availability (through rain-water harvesting) and reduce the demand through water conservation options as well as re-use options.

The proposed alternatives in the industrial sectors proved to be financially feasible on the basis of the existing water price. In the domestic sector the proposed methods were financially infeasible because of the high investment required for the new interventions. However, the social benefits gained from improved health and social life, not included in the calculations, may justify these investments. In the agricultural sector the proposed methods were financially infeasible too because of the existing low prices for agriculture. Increasing water prices in the agricultural sector will motivate farmers to use treated used-water for irrigation.

Legislations and laws regarding the introduction of these alternatives is an important supporting tool. Awareness and education about water scarcity and potential methods for dealing with it is crucial to achieve effective management.

Implementing the combination of water management measures as proposed in this thesis will put the water use and management on the sustainability track.

7.2 Recommendations for further research

1. The water footprint as defined in this thesis consists of two components. The consumption component which refers to the total volume of fresh water used to produce goods consumed by an individual, business or nation plus the contamination component which refers to the volume of fresh water needed to assimilate the waste produced by that individual, business or nation. The contamination component was estimated many times larger than the consumption component making the water footprint many times larger. Further

research is needed in order to assess the contamination component of the water footprint.

2. One of the water management methods proposed in this study is rainwater harvesting. Further research is needed to investigate the effect of rainwater harvesting on groundwater recharge and runoff.

3. Research is needed into the effects of virtual water trade on a country's economy and political stability.

4. Still much is unknown about water use in industries in Palestine. In order to identify potential opportunities for reducing industrial water consumption, more research is urgently required.

Samenvatting

De West Bank in Palestina is gesitueerd in de centrale hooglanden van Palestina. Het gebied wordt gegrensd door de Rivier de Jordaan en het Dode Zee in het oosten en de grens in het noorden, het westen en het zuiden, overeengekomen tijdens het wapenstilstandoverleg van 1948. Het gebied, dat een oppervlakte beslaat van 5.800 km^2, is semi-aride met beperkte watervoorraden; de belangrijkste watervoorraad voor Palestijnen in de Westbank is grondwater. Het watergebruik per hoofd van de bevolking, voor alle doeleinden, is 50 m^3/jaar volgens de welke De West Bank beschouwd kan worden als uiterst waterschaars. Echter, hoewel de Palestijnen nu al nauwelijks aan hun waterbehoeften kunnen voldoen, zal de situatie in de nabije toekomst nog benarder worden als gevolg van de verwachte groei van de bevolking en de ontwikkeling in de sociale, commerciële, industriële en milieusector.

De politieke situatie in het gebied maakt de waterkwestie nog ingewikkelder. Sinds 1967, toen Israël De West Bank bezette, staan de watervoorraden onder controle van het leger van Israel dat het gebruik hiervan door Palestijnen sterk heeft beperkt Zo wordt aan Palestijnen de toegang tot het water van de Jordaan geheel ontzegd. Voorts bestaat er een ongelijke verdeling van de watervoorraden tussen Palestijnen en Israëliërs: per hoofd van de bevolking is het watergebruik door de Israëliërs zes keer zo groot als dat door de Palestijnen. Bovendien is de toekomstige watertoewijzing tussen Palestijnen en Israëliërs onduidelijk.

Bovenop de politiek-geinspireerde waterschaarste, komt ook nog de wijze waarop in De West Bank wordt omgegaan met water. Water wordt na gebruik geloosd zonder dat hergebruik wordt overwogen. In de meeste gevallen wordt het gebruikte water, ook wel afvalwater genoemd, gelost in *wadis* (droge riverbeddingen) zonder enige vorm van behandeling, met als gevolg een verslechtering van de waterkwaliteit en dus een verminderde beschikbaarheid van een goede kwaliteit water. Waterdistributiesystemen lekken en in gebieden waar water beschikbaar is zijn consumptiecijfers hoog ten koste van gebieden waar geen water voorziening is.

De huidige studie is gebaseerd op de veronderstelling dat Israëlische controle van de Palestijnse watervoorraden een feit blijft gedurende de ontwerpperiode van de studie. Dit is het "worst-case" scenario. Echter, mocht meer water beschikbaar komen, dan zal deze de waterschaarste verlichten.

In het kort lijdt De West Bank aan extreme waterschaarste, heeft (om politieke redenen) minder water dan natuurlijk beschikbaar is, volgt de *"gebruik en loos"* benadering en voorziet een verhoging van de vraag naar water om redenen van bevolkings- en de economische groei. Daarom heeft De West Bank dringend behoefte aan een radicale verandering in de manier van denken over water: van de huidige benadering waarbij je water weggooit na gebruik naar een waarbij water gezien wordt als een schaars goed om zodoende te gekomen tot een situatie waarbij huishoudelijke, landbouwkundige en industriële waterbehoeften kunnen worden gedekt met de beperkte beschikbare watervoorraden en milieueffecten van het gebruikte water beduidend worden verminderd. Het doel van dit onderzoek is een kader te ontwikkelen voor het duurzame beheer van watervoorraden in De West Bank. Om dit doel te bereiken werden drie doelstellingen geformuleerd. De eerste doelstelling is een inventarisatie te maken van het huidige watergebruik in De West Bank door het bepalen van de nationale "water footprint". De tweede doelstelling is het evalueren van technische opties, passend binnen de Palestijnse sociale, culturele, godsdienstige en economische voorwaarden, voor het verbeteren van het waterbeheer, nu en in de nabije toekomst, om waterduurzaamheid tegen 2025 te bereiken. Hiervoor werden drie casussen bestudeerd die de haalbaarheid aantonen van optimaal beheer van water voor huishoudelijke, industriële en landbouwkundig doeleinden. De derde doelstelling is de ontwikkeling van een strategie, voor het optimaal beheer van water voor De West Bank, die leidt tot waterduurzaamheid in 2025.

Het huidige watergebruik in De West Bank werd geraamd door het maken van een waterbalans en door het bepalen van de "water footprint" van de Palestijnen in De West Bank. Hierdoor werd gevonden dat de consumptiecomponent van de water "footprint" 1.116 m^3/hoofd/jaar (gemiddelde over 1998 - 2003) is, in vergelijking met het globale gemiddelde van 1.243 m^3/hoofd/jaar (gemiddelde over 1997 - 2002). De lokale watervoorraden leverden 50 m^3/hoofd/jaar waarvan slechts 16 m^3/hoofd/jaar werd gebruikt voor huishoudelijke doeleinden. Dit is slechts 28% van het globale gemiddelde en 21% van het Israëlische huishoudelijke watergebruik. Deze informatie bevestigt opnieuw de behoefte aan de bovengenoemde verschuiving in het denken in termen van 'gebruik en loos' naar een benadering van 'gebruik, behandel en hergebruik'.

De casus met betrekking tot huishoudelijk watergebruik onderzocht opties (gebruik van regenwater, waterarme en waterloze toiletten, ... enz.) voor beter huishoudelijk watergebruik. Deze opties werden vervolgens financieel, milieutechnisch en sociaal geëvalueerd met behulp van Levens Cyclus Analyse. De belangrijkste conclusie was dat door een combinatie van huishoudelijk watergebruik opties te introduceren, een daling van de watergebruik van zo'n 50% kan worden bereikt, als gevolg waarvan de druk op de schaarse watervoorraden vermindert. Naast deze winst voor het milieu worden ook de kosten verminderd. Op het sociale vlak werd duidelijk dat het introduceren van dergelijke opties de 'quality of life' kan verbeteren van hen die geen toegang tot voldoende water van goede kwaliteit hebben. Bovendien kan het huis van morgen grotendeels onafhankelijk zijn in termen van drink- en afvalwater.

De tweede casus, met betrekking tot irrigatie in De West Bank, is gericht op het optimaliseren van het gebruik van het irrigatiewater middels een lineair wiskundig model. Drie scenario's werden bekeken: Het eerste scenario analyseert de irrigatie van de huidige gewasdistributie, het tweede scenario maximaliseert de winst bij beperkte beschikbaarheid van water en land, en het derde scenario maximaliseert de winst bij beperkte beschikbaarheid van water en land en locale consumptie van de gewassen.

De resultaten van de studie toonden aan dat door het bepalen van het optimale distributie van de vijf, voor deze studie gekozen, gewassen, bij beperkingen van land en water, het watergebruik in de gehele agrarische sector kan verminderen met 4% terwijl dit de winst in de sector met ongeveer 4% kan vergroten.

Er kon worden geconcludeerd dat waterschaarste kan worden verminderd door de volgorde van gewaskeuze te bepalen op basis van het watergebruik. Voorts is de uitbreiding van regenafhankelijke landbouw een hoofdthema in de planning van de gewaskeuze in water schaarse landen.

De casus industrieel watergebruik onderzocht de mogelijkheden voor water- en chemicaliënbesparing in het ontharing - kalk proces van de leerindustrie. Er kon worden geconcludeerd dat het afvalwater van dit proces, na gedeeltelijke reiniging, kon worden hergebruikt. Zodoende kon een substantiële vermindering (tot 58%), van watergebruik worden bereikt in combinatie met besparing van kosten voor chemicaliën en vermindering van milieueffecten.

Tot slot werd een strategie voor duurzame waterbeheer in De West Bank ontwikkeld. De bestaande situatie van de watersector werd geanalyseerd in termen van beschikbare watervoorraden, watergebruik, Palestijnse waterrechten, het Nationale Waterplan, de institutionele en organisatorische structuur van de watersector, als ook de verwachte beschikbaarheid en vooruitzichten in zake de vraag naar water tot aan het jaar 2025. Tegen deze achtergrond werden drie scenario's besproken:

- Het "do-nothing" scenario dat veronderstelt dat de bestaande waterbeschikbaarheid niet zal veranderen (Israël handhaaft de controle over de Palestijnse watervoorraden) terwijl de bevolking stijgt waardoor de vraag naar water groeit;
- Het "waterschaarste" scenario dat veronderstelt dat meer water beschikbaar zal komen na succesvolle onderhandelingen tussen Palestijnen en Israëliërs. Desondanks, zullen de bevolkingstoename en de verbeteringen van de sociale, commerciële, industriële en milieusectoren de vraag naar water verhogen;
- Het " duurzame watergebruik" scenario dat, op basis van de resultaten van een SWOT analyse, een pakket aan maatregelen voorstelt op het gebied van technische en institutionele verbeteringen, de vereiste wetgeving en verordeningen voor het realiseren van deze verbeteringen, het nodige onderwijs voor gebruikers en voor hen die deze veranderingen moeten implementeren als ook de noodzakelijke economische motivatie.

Onder zowel het "do-nothing" scenario als ook het "waterschaarste" scenario" is er een groeiende discrepantie tussen waterbeschikbaarheid en de vraag naar water. Echter, de voorgestelde strategie in het "duurzame watergebruik" scenario toont aan dat dit gat kan worden gedicht door het introduceren van opties voor waterbeheer die tot doel hebben om de beschikbaarheid van water te vergroten, door regenwater op te vangen, en de vraag naar water te verminderen, door waterbesparing- en hergebruikopties.

De voorgestelde alternatieven in de industriesectoren bleken financieel uitvoerbaar te zijn op basis van de bestaande waterprijs. In de huishoudelijke watersector bleken de voorgestelde methoden financieel onhaalbaar wegens de vereiste hoge investering. Echter, deze investeringen worden financieel aantrekkelijk wanneer de sociale en economische voordelen van betere gezondheid en sociaal welzijn in de berekeningen worden verdisconteerd. De World Health Organization schat deze voordelen op ruim 8 US$ per geïnvesteerde dollar. Bovendien worden zijn investeringen verondersteld te worden betaald door mensen die van deze verbeteringen gaan profiteren. De bereidheid hiertoe zal groter worden bij toenemende waterschaarste. Het zetten van tarieven die de waterschaarste reflecteren zal mensen motiveren om hierin te investeren. Bovendien ligt het in de lijn der verwachtingen dat de internationale gemeenschap zal bijdragen aan het realiseren van de nodige infrastructuur.

Met de huidige lage waterprijzen in de landbouwsector, is de voorgestelde benadering voor de besparing van water in de landbouw financieel onaantrekkelijk, ook wegens de hoge investeringskosten. Echter, het ontwerpen van een redelijk systeem van tarifering, dat kostenterugwinning in de landbouwsector zeker stelt, zal landbouwers stimuleren tot het gebruiken van behandeld afvalwater voor irrigatie doeleinden.

Het ten uitvoer leggen van het in deze thesis beschreven pakket van maatregelen, zal het beheer van het water in De West Bank in Palestina op het spoor zetten van duurzaamheid.

About the Author

 Dima Wadi' Nazer was born on March 27 1958 in Hebron city in Palestine. In 1981 she was awarded a bachelor degree in civil engineering from the Faculty of Engineering and Technology in the University of Jordan, Amman, Jordan. After graduation she worked as a site engineer for the UNRWA engineering office in Jerusalem, Palestine. In 1982 Dima joined the Jordanian Cooperative Organization as an advising Engineer for the housing cooperatives in the West Bank until 1994 when she was offered a job of a coordinator in the continuing education department in the University Graduate Union in Hebron. In 1995 she joined Palestine Technical College Al-Arroub as the Dean Deputy. In parallel with her work in Palestine Technical College she pursued her study for an M.Sc. program in water engineering in the Institute of Water Studies in Birzeit University from where she was awarded her second degree in 2002. In 2004 she perused her PhD studies with UNESCO-IHE Institute for water education.

Address
Dima Wadi' Nazer
P.O. Box 606
Hebron, West Bank
Palestine
Telefax; +970 (0)22251725
E-mail :   nazerdima@hotmail.com

T - #0105 - 071024 - C40 - 246/174/9 - PB - 9780415573818 - Gloss Lamination